Thin-layer Chromatography
A Modern Practical Approach

RSC Chromatography Monographs

Series Editor: R.M. Smith, *Loughborough University of Technology, UK*

Advisory Panel: J.C. Berridge, *Sandwich, UK*, G.B. Cox, *Indianapolis, USA*, I.S. Lurie, *Virginia, USA*, P.J. Schoenmakers, *Eindhoven, The Netherlands*, C.F. Simpson, *London, UK*, G.G. Wallace, *Wollongong, Australia*

Other titles in this series:

Applications of Solid Phase Microextraction
Edited by J Pawliszyn, *University of Waterloo, Waterloo, Ontario, Canada*

Capillary Electrochromatography
Edited by K D Bartle and P Myers, *University of Leeds, UK*

Chromatographic Integration Methods, Second Edition
N Dyson, *Dyson Instruments, UK*

Cyclodextrins in Chromatography
By T Cserháti and E Forgács, *Hungarian Academy of Sciences, Budapest, Hungary*

Electrochemical Detection in the HPLC of Drugs and Poisons
R J Flanagan, *Guy's and St Thomas' NHS Foundation Trust, London, UK*, D Perrett, *Queen Mary's School of Medicine and Dentistry, London, UK* and R Whelpton, *University of London, London, UK*

HPLC: A Practical Guide
T Hanai, *Health Research Foundation, Kyoto, Japan*

Hyphenated Techniques in Speciation Analysis
Edited by J Szpunar and R Lobinski, *CNRS, Pau, France*

Packed Column SFC
T A Berger, *Hewlett Packard, Wilmington, Delaware, USA*

Separation of Fullerenes by Liquid Chromatography
Edited by *Kiyokatsu Jinno, Toyohashi University of Technology, Japan*

Validation of Chromatography Data Systems: Meeting Business and Regulatory Requirements
R D McDowall, *McDowall Consulting, Bromley, Kent, UK*

How to obtain future titles on publication:

A standing order plan is available for this series. A standing order will bring delivery of each new volume upon publication. For further information please contact:
Sales and Customer Care, Royal Society of Chemistry, Thomas Graham House, Science Park, Milton Road, Cambridge, CB4 0WF
Telephone: +44(0) 1223 420066, Fax: +44(0) 1223426017, Email: sales@rsc.org

RSC CHROMATOGRAPHY MONOGRAPHS

Thin-layer Chromatography
A Modern Practical Approach

Peter E. Wall
VWR International Ltd., Poole, Dorset

advancing the chemical sciences

ISBN 0-85404-535-X

A catalogue record for this book is available from the British Library

Published by The Royal Society of Chemistry,
Thomas Graham House, Science Park, Milton Road,
Cambridge CB4 0WF, UK

Registered Charity Number 207890

For further information see our web site at www.rsc.org

Typeset by Alden Bookset, Northampton, UK
Printed by Athenaeum Press Ltd, Gateshead, Tyne and Wear, UK

Preface

Thin-layer chromatography (TLC) is without doubt one of the most versatile and widely used separation methods in chromatography. Commercially, many sorbents on a variety of backings are now available. Most stages of the technique are now automated (can now be operated instrumentally) and modern HPTLC (High performance thin-layer chromatography) allows the handling of a large number of samples in one chromatographic run. Speed of separation (development time), high sensitivity and good reproducibility all result from the higher quality of chromatographic layers and the continual improvement in instrumentation. In addition TLC has remained relatively inexpensive and one can easily see why it is still popular today. It has found a use in a wide range of application areas as the concept of TLC is so simple and samples usually require only minimal pre-treatment. It is often thought of only in terms of its use in pharmaceutical analysis and production and in clinical analysis, but many standard methods in industrial chemistry, environmental toxicology, food chemistry, water, inorganic and pesticide analysis, dye purity, cosmetics, plant materials, and herbal analysis rely upon TLC as the preferred approach. In its simplest form, TLC costs little, but even including the more sophisticated instrumentation, it still remains less expensive per sample analysis than, for example HPLC. Thin-layer chromatography continues to be an active technique in research with about 500–700 publications appearing each year during the 1980s and 1990s.

It is the purpose of this book to describe the advances made, particularly in the last two to three decades, which have revolutionised TLC and transformed it into a modern instrumental technique. All aspects of TLC have been affected, from the sorbent layer technology, through the "spotting" devices and developing equipment, to the final detection and quantification. Computers too, now play an important role in the control of equipment and in the computation of the vast amount of data obtained from scans of the developed TLC layer. This has resulted in the ability to store and retrieve images of chromatograms and physical data on actual separation results and conditions for future use. Instrumental planar chromatography is now capable of handling samples with minimal pre-treatment, detecting components at low nanogram sensitivities and with relative standard deviations of about 1%. It is the opinion of the author that these developments demonstrate that the previous image of low sensitivity, poor resolution and reproducibility can be discarded and that TLC is now truly a modern contemporary of HPLC and GC. Modern TLC has become a powerful, reliable and cost efficient method for qualitative and quantitative analysis.

The chapters in this book have been designed in such a way that the reader follows each step of the planar chromatographic process in logical order. Hence the choice of sorbent is followed by preparation of sample for application and the methods of application. The subject of chromatogram development logically comes next with detection, quantification and/or video imaging usually being the final steps. However, sometimes further instrumental analysis is necessary, hence the final chapter on hyphenated techniques. Any theory or necessary mathematical equations or expressions are introduced when required within the text of each subject. As the objective of this book is to provide a publication or manual that can be used by the practising chromatographer, the depth of theory reflects only what is required to explain why certain practical steps are taken. It is the intention of the author that this book will be of practical value and use to those who are contemplating using TLC for the first time, and also to those who have been planar chromatographers for some time. With these points in mind, the practical examples of chromatographic separations reflect the field-tested procedures available. The book concentrates on the basic steps involved in TLC, providing practical guidance to achieve superior separations on a TLC/HPTLC sorbent layer. For this reason there are some techniques that are not covered, such as sorbent coated quartz rods and thin-layer radiochromatography. However, the basic principles for optimised separations described in the various chapters will still apply in these related techniques. Recommendations made throughout the text to obtain acceptable and often high quality results are made on the basis of many years of practical experience in planar chromatography by the author.

Numerous commercial products are referred to in this book as would be necessary with any publication that discusses instrumental TLC/HPTLC. The references made are based on the experiences of the author with these products, and are definitely not meant to imply that they are superior to comparable products from other manufacturers.

Contents

Chapter 7   Quantification and Video Imaging   154

Chapter 8   TLC Coupling Techniques   166

Subject Index   177

CHAPTER 1

Introduction and History

1 Introduction to Thin-layer Chromatography

The basic TLC procedure has largely remained unchanged over the last fifty years. It involves the use of a thin, even sorbent layer, usually about 0.10 to 0.25 mm thick, applied to a firm backing of glass, aluminium or plastic sheet to act as a support. Of the three, glass has always proved the most popular, although aluminium and plastic offer the advantage that they are flexible and can more easily be cut to any size with minimal disruption to the sorbent layer. Numerous sorbents have been used, some more successfully than others, including silica gel, cellulose, aluminium oxide, polyamide and chemically bonded silica gels. The sample is dissolved in an appropriate solvent and applied as spots or bands along one side of the sorbent layer approximately 1 cm from the edge. An eluent (single solvent or solvent mixture) is allowed to flow by capillary action through the sorbent starting at a point just below the applied samples. Most commonly this is achieved by using a glass rectangular tank in which the eluent is poured to give a depth of about 5 mm. The plate is placed in the tank or chromatography chamber and the whole covered with a lid. As the eluent front migrates through the sorbent, the components of the sample also migrate, but at different rates, resulting in separation. When the solvent front has reached a point near the top of the sorbent layer, the plate or sheet is removed and dried. The spots or bands on the developed layer are visualised, if required, under UV light or by chemical treatment or derivatisation. For quantitative determinations, zones can be removed or eluted from the layer, or the plate can be scanned at pre-determined wavelengths without disturbing the layer surface. The modern use of TLC has seen a strong move in the direction of plate scanning and video imaging as a means of providing sensitive and reliably accurate results and a more permanent record of the chromatogram. This is in addition to its obvious labour saving aspect and chemically "clean" approach.

Although TLC is an analytical method in its own right, it is also complimentary to other chromatographic techniques and spectroscopic procedures. Results obtained with TLC can often be transferred to HPLC or vice versa with some adjustment in eluting solvent conditions. For multi-component samples (*e.g.* pesticides in water), fractions of interest from an HPLC separation can be collected and subsequent re-chromatography of these on HPTLC can give a "fine tuned" separation of the components of the fractions.[1-3] Thin-layer chromatography has

1

been successfully hyphenated with high performance liquid chromatography (HPLC), mass spectroscopy (MS), Fourier transform infra-red (FTIR), and Raman spectroscopy, to give far more detailed analytical data on separated compounds. Even the UV/visible diode array technique has been utilised in TLC to determine peak purity or the presence of unresolved analytes.[4]

Undoubtedly TLC is a modern analytical separation method with extensive versatility, much already utilised, but still with great potential for future development into areas where research apparently is only just beginning.

2 History of TLC

Although column chromatography can be traced to its discoverer, the Russian botanist, Tswett in 1903,[5] it was not until 1938 that separations on thin-layers were achieved when Izmailov and Shraiber,[6] looking for a simpler technique, which required less sample and sorbent, separated plant extracts using aluminium oxide spread on a glass plate. The sorbent was applied to a microscope slide as a slurry, giving a layer about 2 mm thick. The sample (plant extracts) was applied as droplets to the layer. The solvent (methanol) was then added dropwise from above on to the applied spots and a series of circular rings were obtained of differing colours on the layer. Circular TLC was born, and Izmailov and Shraiber named this new technique "drop chromatography".

In 1949 Meinhard and Hall used a starch binder to give some firmness to the layer, in order to separate inorganic ions, which they described as "surface chromatography".[7] Further advances were made in 1951 by Kirchner *et al.*,[8] who used the now conventional ascending method, with a sorbent composed of silicic acid, for the separation of terpene derivatives, describing the plates used as "chromatostrips". In 1954, Reitsema[9] used much broader plates and was able to separate several mixtures in one run. Surprisingly it was some time before the advantages of this development were recognised.

However, from 1956 a series of papers from Stahl[10–13] appeared in the literature introducing "thin-layer chromatography" as an analytical procedure, describing the equipment and characterisation of sorbents for plate preparation. Silica gel "nach Stahl" or "according to Stahl" became well known, with plaster of Paris (calcium sulphate) being used as a binder and TLC began to be widely used. In 1962, Kurt Randerath's book on TLC was published,[14] followed by those of Stahl and co-workers, entitled 'Thin-Layer Chromatography – A Laboratory Handbook' (1965),[15] and Kirchner's, 'Thin-Layer Chromatography' (1967).[16] Then, in 1969 a 2nd edition of Stahl's book appeared which was greatly expanded.[17] These authors showed the wide versatility of TLC and its applicability to a large spectrum of separation problems and also illustrated how quickly the technique had gained acceptance throughout the world. (By 1965 Stahl could quote over 4500 publications.)[18] With Stahl's publication the importance of factors such as controlling the layer thickness, layer uniformity, the binder level and the standardisation of the sorbents as regards pore size and volume, the specific surface area and particle size, were recognised as crucial to obtaining highly reproducible, quality separations.

Commercialisation of the technique began in 1965 with the first pre-coated TLC plates and sheets being offered for sale. TLC quickly became very popular with about 400–500 publications per year appearing in the late 1960s as it became recognised as a quick, relatively inexpensive procedure for the separation of a wide range of sample mixtures. As the range and reliability of commercial plates/sheets improved, standard methods for analysis appeared throughout industry. It soon became evident that the most useful of the sorbents was silica gel, particularly with an average pore size of 60 Å, and it was on this material that the commercial companies centred their attention. Modifications to the silica gel began with silanisation to produce reversed-phase layers. This opened up a far larger range of separation possibilities based on a partition mechanism, compared with adsorption as used in most previous methods.

Up to this time quantitative TLC was fraught with experimental error. However, the introduction of commercial spectrodensitometric scanners enabled the quantification of analytes directly on the TLC layer. Initially peak areas were measured manually, but later integrators achieved this automatically.

The next major advance was the advent of HPTLC (High performance thin-layer chromatography). In 1973 Halpaap was one of the first to recognise the advantage of using a smaller average particle size of silica gel (about 5–6 μm) in the preparation of TLC plates. He compared the effect of particle size on development time, R_f values and plate height.[19] By the mid 1970s it was recognised that HPTLC added a new dimension to TLC as it was demonstrated that precision could be improved ten-fold, analysis time could be reduced by a similar factor, less mobile phase was required, and the development distances on the layers could be reduced.[20] The technique could now be made fully instrumental to give accuracy comparable with HPLC. Commercially the plates were first called "nano-TLC" plates by the manufacturer, (Merck[a]), but this was soon changed to the designation "HPTLC". In 1977 the first major HPTLC publication appeared, simply called "HPTLC high performance thin-layer chromatography" edited by Zlatkis and Kaiser.[21] In this volume Halpaap and Ripphahn described their comparative results with the new 5×5 cm HPTLC plates versus conventional TLC for a series of lipophilic dyes.[22] Bonded phases then followed in quick succession.

Reversed-phase HPTLC was reported in 1980 by Halpaap *et al.*[23] and this soon became commercially available as pre-coated plates. In 1982 Jost and Hauck[24] reported an amino (NH_2–) modified HPTLC plate, which was soon followed by cyano-bonded (1985)[25] and diol-bonded (1987)[26] phases. The 1980s also saw improvements in spectrodensitometric scanners with full computer control becoming possible, including options for peak purity and the measurement of full UV/visible spectra for all separated components. Automated multiple development (AMD) made its appearance in 1984 due to the pioneering work of Burger.[27] This improvement enabled a marked increase in number and resolution of the separated components.

In recent years TLC/HPTLC research has entered the chiral separation field using a number of chiral selectors and chiral stationary phases. Only one type of chiral pre-coated plate is presently commercially available, which is based on a

[a] Merck KGaA, Darmstadt, Germany.

ligand exchange principle and is produced commercially either as a TLC or HPTLC plate. Günther has reported results with amino-acids and derivatives on the TLC plate[28] and Mack and Hauck similarly with their HPTLC equivalent.[29]

At the present time all steps of the TLC process can be computer controlled. The use of highly sensitive charge coupled device (CCD) cameras has enabled the chromatographer to electronically store images of chromatograms for future use (identity or stability testing) and for direct entry into reports at a later date. Commercially available HPTLC plates coated with specially pure 4–5 μm spherical silica gel have added further capabilities to the technique. Background interference has been reduced, and resolution further improved, which has enabled TLC to be hyphenated effectively with Raman spectroscopy.

3 References

1. K. Burger, 'Instrumental Thin-Layer Chromatography/Planar Chromatography', *Proceedings of the International Symposium*, Brighton, UK, 1989, 33–44.
2. R.E. Kaiser, 'Instrumental Thin-Layer Chromatography/Planar Chromatography', *Proceedings of the International Symposium*, Brighton, UK, 1989, 251–262.
3. D. Jänchen and H.J. Issaq, *J. Liq. Chromatogr.*, 1988, **11**, 1941–1965.
4. D. Jänchen in *Handbook of Thin Layer Chromatography*, 2nd edn, J. Sherma and B. Fried (eds), Marcel Dekker Inc., New York, USA, 1996, 144.
5. M. Tswett, *Proc. Warsaw Soc. Nat. Sci., Biol. Section*, 1903, **14**, minute no. 6.
6. N.A. Izmailov and M.S. Shraiber, *Farmatisiya*, 1938, no. 3, 1.
7. J.E. Meinhard and N.F. Hall, *Anal. Chem.*, 1949, **21**, 185.
8. J.G. Kirchner, J.M. Miller and G.E. Keller, *Anal. Chem.*, 1951, **23**, 420.
9. R.H. Reitsema, *Anal. Chem.*, 1954, **26**, 960.
10. E. Stahl, *Pharmazie*, 1956, **11**, 633.
11. E. Stahl, *Chemiker-Ztg*, 1958, **82**, 323.
12. E. Stahl, *Pharmaz. Rdsch.*, 1959, **2**, 1.
13. E. Stahl, *Angew. Chem.*, 1961, **73**, 646.
14. K. Randerath in *Thin-Layer Chromatography*, Academic Press, London, UK, 1963.
15. *Thin-Layer Chromatography – A Laboratory Handbook*, E. Stahl (ed), Springer-Verlag, Berlin, Germany, 1965.
16. J.G. Kirchner, *Thin-Layer Chromatography*, 2nd edn, Techniques in Chemistry, vol. XIV, Wiley-Interscience, Chichester, UK, 1978.
17. *Thin-Layer Chromatography – A Laboratory Handbook*, 2nd edn, E. Stahl (ed), Springer-Verlag, Berlin, Germany, 1969.
18. E. Stahl, in *Thin-Layer Chromatography*, 2nd edn, E. Stahl (ed), Springer-Verlag, Berlin, Germany, 1969, 5.
19. H. Halpaap, *J. Chromatogr.*, 1973, **78**, 77–78.
20. A. Zlatkis, R.E. Kaiser, in *HPTLC high performance thin-layer chromatography*, A. Zlatkis, and R.E. Kaiser (eds), Elsevier, Amsterdam, Netherlands, 1977, 12.

21. *HPTLC high performance thin-layer chromatography*, A. Zlatkis and R. E. Kaiser (eds), Elsevier, Amsterdam, Netherlands, 1977.
22. H. Halpaap and J. Ripphahn, in *HPTLC high performance thin-layer chromatography*, A. Zlatkis and R.E. Kaiser (eds), Elsevier, Amsterdam, Netherlands, 1977, 95–125.
23. H. Halpaap, K.F. Krebs and H.E. Hauck, *J. HRC and CC*, 1980, **3**, 215–240.
24. H.E. Hauck and W. Jost, 'Instrumental High Performance, Thin-Layer Chromatography', *Proceedings of 2nd International Symposium*, R.E. Kaiser (ed), Interlaken, Switzerland, 1982, 25–37.
25. H.E. Hauck and W. Jost, 'Instrumental High Performance Thin-Layer Chromatography', *Proceedings of 3rd International Symposium*, R.E. Kaiser (ed), Würzburg, Germany, 1985, 83–91.
26. H.E. Hauck and W. Jost, 'Instrumental High Performance Thin-Layer Chromatography', *Proceedings of 4th International Symposium*, R.E. Kaiser, H. Traitler and A. Studer (eds), Selvino/Bergamo, Italy, 1987, 241–253.
27. K. Burger, *Z. Anal. Chem.*, 1984, **318**, 228.
28. K.J. Günther, *J. Chromatogr.*, 1988, **448**, 11–30.
29. M. Mack, H.E. Hauck and H. Herbert, *J. Planar Chromatogr.*, 1988, **1**, 304–308.

CHAPTER 2

Sorbents and TLC Layers

1 Sorbent Selection

1.1 Introduction

There are at least 25 inert materials that are available as sorbents in TLC, some of which have been more widely used than others. A number of the more important ones will be reviewed in this chapter. Clearly for optimum separations, it is important that the correct material is chosen. Some sorbents have a specific range of application (*e.g.* silica gel impregnated with caffeine for polyaromatic hydrocarbons, or silica gel impregnated with a chiral selector for the separation of enantiomers of amino-acids and derivatives). By contrast silica gel or aluminium oxide are used for a wide range of applications. Silica gels and aluminas can also be split into a number of distinct, separate sorbents depending on pore size, particle size, and pH. Before choosing the sorbent, consideration must be given to the compounds to be separated. Characteristics, such as the polarity, solubility, ionisability, molecular weight, shape and size of the analytes are all important in deciding on a separation mechanism, and hence largely define both the type of sorbent and the solvents used both for the preparation of the sample and in development.

In 1973 Scott[1] examined over 1100 papers to determine which sorbents were the most regularly used in TLC. Silica gel was by far the most popular (~64%), followed by cellulose (~9%), and alumina (~3%). Since then silica gel has remained the most widely used, but noticeable changes have occurred with the appearance of chemically bonded phases which have opened up a new range of separation possibilities. The newer stationary phases have tended for the most part to address specific areas of separation where either the resolution of sample components was poor or non-existent. As the lists of applications for some sorbents is extensive, it is better to refer to the excellent bibliographies or abstract services that are available for TLC (*e.g.* Camag Bibliography Service[a]) when a specific method from the literature is required. If this is not available to the user or a new or improved procedure is needed, then the basic information in Table 1 will be of help to ensure that the optimum sorbent for the type of separation is chosen.

---

[a] CAMAG, Muttenz, Switzerland.

6

Table 1 *Choice of optimum TLC/HPTLC sorbents for compounds and compound classes*

Sorbent	Compounds separated
Silica gel	All classes of compounds.
Aluminium oxide	Basic compounds (alkaloids, amines, *etc.*), steroids, terpenes, aromatic and aliphatic hydrocarbons.
Cellulose	Amino-acids and derivatives, food dyes (acidic and basic), carbohydrates.
Kieselguhr	Carbohydrates, aflatoxins, herbicides, tetracyclines.
Polyamide	Phenols, flavonoids, nitro-compounds.
Amino-bonded silica gel	Particularly good for carbohydrates, sulfonic acids, phenols, carboxylic acids, nucleotides, nucleosides.
Cyano-bonded silica gel	Many classes of compounds, particularly good for pesticides, steroids, preservatives.
Diol-bonded silica gel	Many classes of compounds, particularly good for steroids, hormones.
Reversed-phase (RP 2, RP 8, RP 18) silica gel	Improves separation for many classes of compounds (*cf.* silica gel) – steroids, tetracyclines, phthalates, antioxidants, lipids, barbiturates, capsaicins, aminophenols, fatty acids.
Chiral (CHIR) modified silica gel	Enantiomers of amino-acids, halogenated, N-alkyl, and α-methyl amino-acids, simple peptides, α-hydroxycarboxylic acids (catecholamines).
Silica gel impregnated with silver nitrate	Lipids, including variations in unsaturation and geometric isomers.
Silica gel impregnated with caffeine	Particularly selective for polyaromatic hydrocarbons.
Silica gel impregnated with boric acid/phosphate	Particularly selective for carbohydrates.

1.2 Silica Based Sorbents

1.2.1 Silica Gel

Silica gel, also called silicic acid and kieselgel, is a white amorphous porous material, usually made by precipitation from silicate solutions by addition of acid. The process is by no means simple as polysilicic acids are formed by polycondensation. So-called "primary particles" appear. As the particles grow, water is eliminated and gel formation takes place. The control of the temperature and pH during this stage will have a marked bearing on the quality of the gel formed. Colloidal particles thus develop, which further condense and shrink to form a three dimensional network described as a hydrogel[2] (see Figure 1). After washing

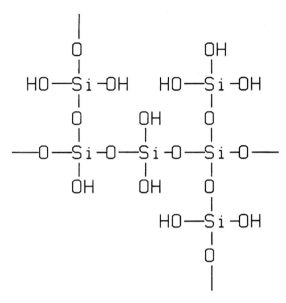

Figure 1 *Section of typical silica gel structure*

and heating (~120 °C) an amorphous hard but porous gel is formed, called a xerogel or silica gel. It is this xerogel that is used for TLC.

The structure is held together by bonded silicon and oxygen, termed siloxane groups. Residual hydroxyl groups on the surface account for much of the adsorptive properties of silica gel giving it unique separation characteristics. These "active sites" can vary according to their local environment. Three types of hydroxyl group are possible as shown in Figure 2. The most prolific is the single hydroxyl group bonded to a silicon atom, which is linked to the silica gel matrix via three siloxane bonds. The second type is where two hydroxyl groups are bonded to a single silicon atom, often called a geminal hydroxyl group. The third type, which is much more rare, is a bonding of three hydroxyl groups to one silicon atom. Only a single siloxane bond binds this group to the silica gel matrix.[3]

However, the available "active" surface is somewhat more complicated by the presence of water that hydrogen bonds to the surface hydroxyl groups. Figure 3 shows that there are quite a number of different ways hydrogen-bonding can occur with water. Even multi-layers of water, physically adsorbed, are possible.

The hydration of the gel for TLC is considered to be 11–12% water when the relative humidity is 50% at 20 °C.[4] Such a gel is normally ready for use requiring no pre-activation. Activation is only necessary if the TLC plate has been exposed to high humidity, and then only requires heating up to 105 °C for 30 minutes followed by cooling in a clean atmosphere at 40–50% relative humidity. The water is held in the structure either as physically adsorbed or hydrogen-bonded water, the latter being more firmly held.[5] As proof of this, desorption of hydrogen-bonded water requires 10 kcal mol^{-1} whereas physically bound water requires, 6.6–8.2 kcal mol^{-1} activation energy.[6]

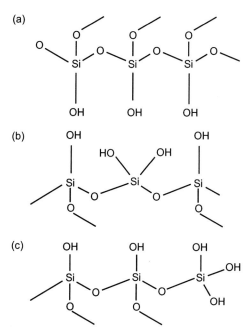

Figure 2 *Silica gel surface hydroxyl group types* (a) *monohydroxyl (single),* (b) *dihydroxyl (geminal),* (c) *trihydroxyl*

The synthetic nature of silica gel for chromatography enables the careful control of pore size, pore volume and particle size. Pore size varies from 40 to 150 Å for commercial pre-coated TLC plates with one notable exception of 50 000 Å for special applications. The range of particle size of silica gel for TLC is typically 5 to 40 μm with the average being 10 to 15 μm depending on the manufacturer. This has a large effect on the resolution of sample components. Thus in TLC, as in HPLC, reducing the particle size lowers the height equivalent to a theoretical plate of a peak and hence increases the efficiency. As illustrated in Figure 4, when smaller silica gel particles of 5 to 6 μm are used to prepare HPTLC plates, improved resolution results.

Pore size affects selectivity and hence can be used to good effect in altering the migration rates and resolution of sample components. The most common pore sizes used in TLC are 40, 60, 80, 100 Å, with silica gel 60 Å being by far the most popular and versatile of the group. Silica gel 60 Å (commonly called silica gel 60) has been recommended for a wide range of separations throughout industry and research institutions. As water content plays such an important role in the retention of analytes on the chromatographic layer, it is vital that the moisture adsorbed by the silica gel is maintained at a constant level. In Figure 5 water adsorption curves are shown for the range of silica gel pore sizes; 40, 60 and 100 Å. At normal levels of humidity in most laboratories (40–60% relative humidity), the variation in uptake of moisture by silica gel 60 Å has little effect

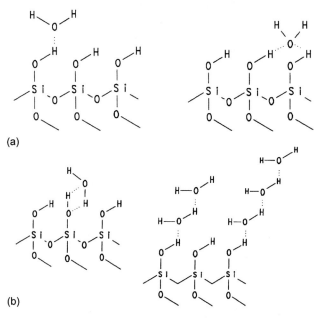

Figure 3 (a) *Ways in which water hydrogen bonds to the surface hydroxyl groups of silica gel; (b) Formation of multi-layers of hydrogen bonded water*

on the migration rates of most sample components. The change in water adsorption over small changes in relative humidity (RH) for silica gel 40 Å is quite marked (from 20–40% over an RH range of 40–60%). This will affect migration rates of the sample components, and although with careful control

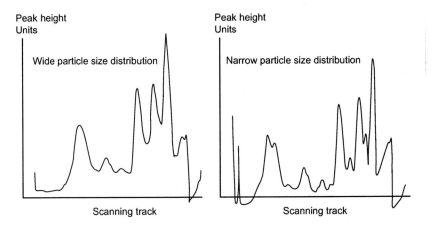

Figure 4 *Effect of particle size distribution on resolution. Separation of carotene concentrate on silica gel 60*

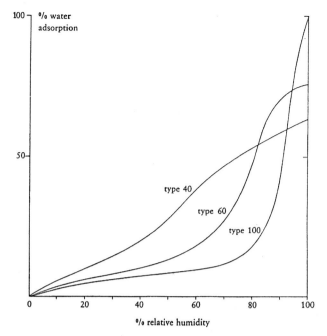

Figure 5 *Water adsorption isotherms for silica gel 40, 60 and 100* Å
(By permission of Merck)

humidity differences can be utilised to improve separations, they can also be a source of problems with respect to reproducibility. Although humidity control is not of so much concern for silica gel 100 Å, the sorbent is consequently less polar chromatographically due to comparatively less moisture adsorption resulting in low migration rates of sample components.

Because variations in water content with relative humidity obviously exist, Brockmann and Schodder introduced a scale to characterise various silica gels (see Table 2). This scale graded silica gels from I to V. It was designed to characterise silica gels according to their chromatographic activity or selectivity as the adsorbed water changes. As the water content increases, the chromatographic layer becomes

Table 2 *Brockmann and Schodder activity grading of silica gels according to water content*[62]

Relative humidity	0%	20%	40%	60%	80%
Silica gel 40 Å	I	II	III	IV–V	>V
Silica gel 60 Å	I–II	II	III	III–IV	V
Silica gel 100 Å	II	II–III	III	III–IV	IV

Table 3 *Variation of physical characteristics of silica gels according to pore diameter*

Sorbent	Pore volume (ml g^{-1})	Specific surface area (BET) (m^2 g^{-1})	pH of 10% w/v aqueous suspension
Silica gel 40	0.65	650	5.5
Silica gel 60	0.75	500	7.0
Silica gel 100	1.00	400	7.0

more polar and the solutes applied to the layer show increased migration toward the solvent front even though no change has been made to the solvent in the TLC tank.

Changes in pore diameter will also cause a change in selectivity. At constant relative humidity the uptake of moisture will vary dependent on the pore size as can be seen in Figure 5. For example at 50% relative humidity, the following values can be read from the graph for silica gel with 40, 60 and 100 Å pores; 27%, 13% and 8%, respectively. As a general rule solutes migrate faster with silica gel 40 plates and slower with silica gel 100 compared with silica gel 60. This is an effect of the variation of the polarity of the different types of silica gel. Further physical characteristics of silica gel 40, 60 and 100 are shown in Table 3.

Silica gel TLC plates are extremely versatile over a wide range of applications. Solvent mixtures composed of non-polar (*e.g.* hexane or cyclohexane) and polar (*e.g.* methanol, acetonitrile or water) constituents can be used without the chromatographic layer or binder being affected. (Typical example in Figure 6.) Often acid modifiers (*e.g.* acetic, propionic or formic acids) or base modifiers (*e.g.* ammonia solution, pyridine or amines) are incorporated into the developing solvent to improve resolution. These are also well tolerated by the layers resulting in separations of which Figures 7, and 8 are typical examples.

1.2.2 Silica Gel Bonded Phases

Reversed-Phase. Traditionally silicone or paraffin oils have been used to produce reversed-phase layers. These plates have the advantage of being usable with eluents up to 100% water and are relatively easy to prepare.[7–10] However, they can suffer from stationary phase leakage or "stripping" during chromatography. Often for qualitative work this disadvantage can be accommodated. Where quantitative analysis really matters it is important that the reversed-phase layer is completely reproducible, gives similar and low background interference during scanning and does not alter the mobile phase polarity. This has resulted in the development of bonded-silica gel TLC and HPTLC phases for reversed-phase work.

Silica gel can be chemically bonded by reaction with organosilanes of various chain lengths. Dimethyl, ethyl, octyl, undecyl, octadecyl and phenyl have all been used commercially to produce carbon chain lengths of 2, 8, 12, 18 and aromatic

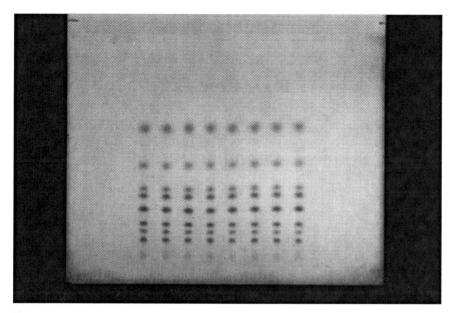

Figure 6 *Separation of corticosteroids on silica gel 60*
Sorbent layer: HPTLC silica gel 60 F_{254} glass plates (Merck)
Mobile phase: chloroform/methanol (93 + 7 v/v)
Detection: 0.5% w/v blue tetrazolium in methanol
Zones: 1, prednisolone; 2, hydrocortisone; 3, prednisone; 4, cortisone; 5, cortico-
sterone; 6, cortexolone; 7, 11-dehydrocorticosterone; 8, 11-desoxycorticosterone
(from lowest R_f)
Concentration: 20 ng per substance

rings bonded to the siloxane matrix. (For an example of the structure, see Figure 9.) All such chemical modifications in TLC and HPTLC format plates are widely available from the major manufacturers of chromatographic layers. The bonding of the organosilanes to the silica gel can be carried out under anhydrous conditions, where a monolayer type bonding, or under hydrous conditions, where a polymer layer occurs. In monolayer formation, mono, bi- or tri-functional organosilanes can be used and the possible reactions with surface silanols are shown in Figure 10. The reaction stoichiometry obviously depends on the concentration of silanol groups (Si-OH) at the surface, and as can be seen from the reaction equations, results in the organo-group being bonded via a siloxane group (Si–O–Si). This is true whether the modification results in a mono or polymeric layer. In the formation of polymeric layers the presence of water initiates the conversion of organosilanes to organosilanetriols by hydrolysis. These immediately undergo condensation with the surface silanols of the silica gel resulting in multiple bonding to the surface.[11]

With bonded phases, only the accessible silanol groups on the silica gel matrix can be modified. The types and degrees of modification result in a marked difference in hydrophobicity between the sorbents. In partition chromatography of this sort where the mobile phase and the stationary phase have been reversed in

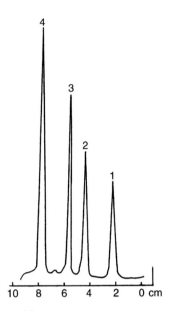

Figure 7   *Separation of flavonoids*
Sorbent layer: HPTLC silica gel 60 glass plates (Merck)
Mobile phase: ethyl acetate/water/formic acid (85 + 15 + 10 v/v)
Detection: 1% w/v diphenylboric acid-2-aminoethyl ester in methanol
Peaks: 1, rutin; 2, hyperoside; 3, quercitrin; 4, quercitin
Concentration: 30 ng per substance
Scanning: fluorescence spectrodensitometry, excitation at 436 nm and emission
at 546 nm. (For details on spectrodensitometry, see Chapter 7.)
Reprinted from *Thin-Layer Chromatography, Reagents and Detection Methods,*
volume 1a, 1990, p 279, by courtesy of the authors and the publisher, Wiley-VCH

polarity, it is imperative that the eluents used are polar, *e.g.* acetonitrile/water or
methanol/water, but not so polar that the surface of the bonded phase remains un-
wetted. As the degree of surface modification and alkyl chain length increases, the
layer becomes more hydrophobic, and it is only possible to use comparatively low
concentrations of water as pressure is not normally applied. Thus some
commercially available RP18 HPTLC plates where the silica gel has been silanised
as fully as possible can only be used with up to 25% water in the developing solvent
mixture. The result of a further increase in water concentration is an uneven solvent
front and incomplete wetting of the layer. However, by reducing the degree of
surface coverage of C18 it is possible to produce completely water tolerant phases,
(*e.g.* the RP18W HPTLC plate from Merck). Logically TLC plates of lower
silanisation can also be produced which are totally wettable. Table 4 shows the
surface coverage attained by one manufacturer. Apart from the fully silanised
HPTLC RP2, 8 and 18, all of the others may be used with high concentrations of
aqueous solvents if required.
    Much has been published on the comparison of different manufacturers' pre-
coated chemically bonded HPTLC/TLC plates and not surprisingly, based on the

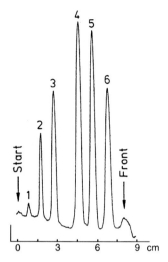

Figure 8 *Separation of alkaloids*
Sorbent layer: silica gel 60 WF₂₅₄ glass plates (Merck)
Mobile phase: acetone/toluene/ethanol/ammonia soln. (25% w/w) (40 + 40 + 6 + 2 v/v)
Visualisation: 0.15% w/v hexachloroplatinic (IV) acid in 3% w/v potassium iodide solution
Peaks: 1, narceine; 2, morphine; 3, codeine; 4, thebaine; 5, papaverine; 6, narcotine
Concentration: 1 μg per substance
Detection: Reflectance spectrodensitometry at 540 nm
Reprinted from *Thin-Layer Chromatography, Reagents and Detection Methods*, volume 1b, 1994, p363, by courtesy of Wiley-VCH

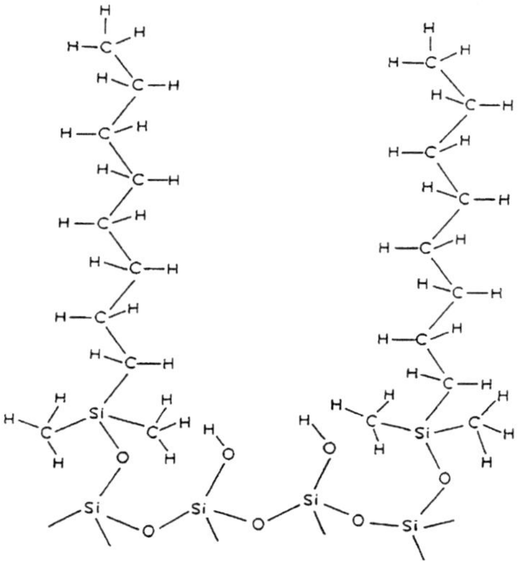

Figure 9 *Chemically bonded C₈ alkyl groups bonded to the silica gel surface*

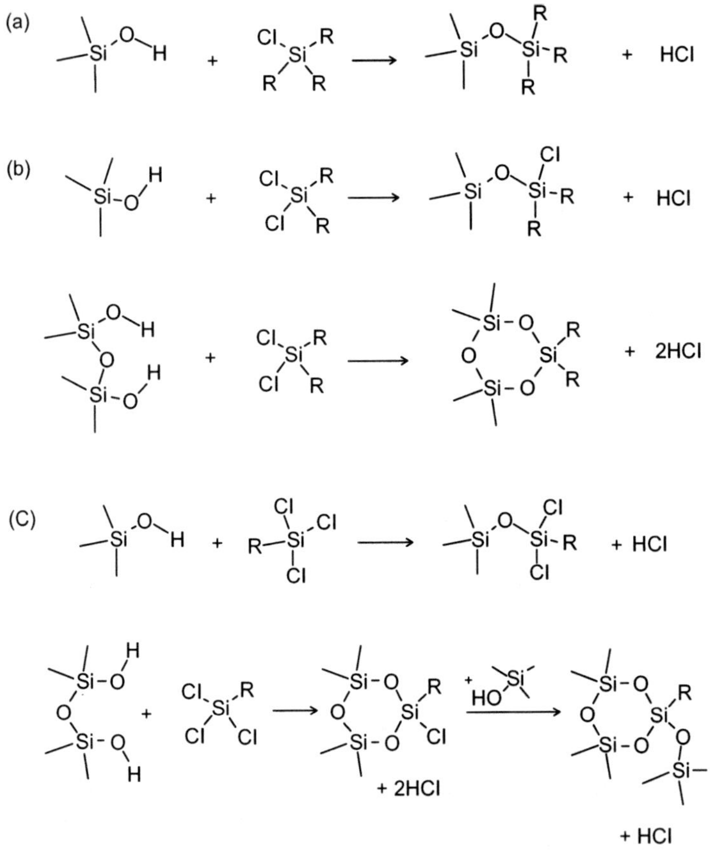

(a)

(b)

(c)

Figure 10 *Reaction between silica gel and* (a) *monofunctional,* (b) *bifunctional, and* (c) *trifunctional organosilanes*

Table 4 *Coverage ratio of chemically modified chromatographic layers*
(By permission of Merck)

Stationary phase	HPTLC (μmol carbon m^{-2})	TLC (μmol carbon m^{-2})
RP 2	3.9	2.5
RP 8	3.0	2.0
RP 18	2.6	1.7
RP 18W (water tolerant layer)	0.6	
NH$_2$	2.9	
CN	3.5	

similarity to HPLC packings, there are noticeable variations.[12–16] For example, some manufacturers recommend the use of up to 3% sodium chloride in the mobile phase to attain better wettability, whereas others do not. In some cases it is not just a matter of the completeness of silanisation of the layer, but also the choice of binder. Many binders are either partially or completely soluble in water. This can result in removal of the sorbent layer with solvents containing a high proportion of water. As these variations can have a large impact on the choice of reversed-phase layer and on the eventual separation quality, it is important to follow the manufacturer's recommendations to obtain the optimum results with their particular plates.

Reversed-phase TLC is now widely used in a variety of applications. To name but a few, Proctor and Horobin have published extensively on its applicability to dye identification and purity in biological stains,[17] Kaiser and Rieder for optimisation of environmental trace analysis,[18] Armstrong *et al.* on polymer molecular weight determinations and distributions,[19,20] Sherma *et al.* on amino-acid separations,[21] Vanhaelen and Vanhaelen-Fastré with the resolution of natural flavonoids and phenols,[22] Amidzhin *et al.* and McSavage and Wall for the densitometric identification of triglycerides,[23,24] and Giron and Groell who have described the separation of diastereoisomers of zeranol.[25] Figures 11 and 12 show two examples which demonstrate the versatility of reversed-phase planar chromatography.

Before leaving the subject of reversed-phase plates, special mention should be made of diphenyl-bonded layers. Although the bonded moiety is aromatic and would be expected to give quite different selectivity to aliphatic bonded phases, it appears to have very similar separation properties to RP2. This plate has been used for the separation of steroids, sulphonamides and peptides, but in most instances the resolutions achieved are little better than those obtained with an ethyl (C2) bonded silica gel layer.[26]

Amino-Bonded Phase. In the same way that various alkyl-bonded sorbents have been prepared, silica gel 60 has also been chemically bonded with aminopropyl groups via siloxane linkages (see Figure 13). The resulting layer is quite stable, and unlike many alkyl-bonded phases is hydrophilic. With careful choice of binder, these plates can be used with mobile phases composed totally of water or buffers. The propyl group has some hydrophobic properties and can therefore be used for separations of a reversed-phase type using aqueous based eluents. Since the amino (-NH$_2$) group is polar, normal-phase chromatography is also possible using much less polar organic eluents than for silica and this is indeed the area where the bulk of the applications have been reported. The phase behaves in many ways like a deactivated silica. The major application for such normal-phase separations has been in the resolution and determination of steroids.

However, it is important to remember that the amino function is a primary amine and hence is chemically quite reactive. Depending on the sample and mobile phase conditions, it is possible to have unwanted reactions occurring on the plate during development, *e.g.* ketones or aldehydes may react with primary amines under alkaline conditions to form Schiff bases. On the other hand, the reactivity of the amino group can be used to advantage. Thus optical isomers can be readily bonded

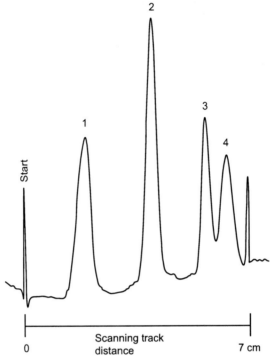

Figure 11   *Separation of nucleobases on a silica gel RP2 HPTLC plate*
Mobile phase: acetonitrile/water (60 + 40 v/v)
Sample applied: 200 nl (0.1% w/v solution)
Detection: UV 254 nm (TLC/HPTLC scanner)
Peaks: 1, cytosine; 2, adenine; 3, guanine; 4, uracil
(By permission of Merck)

under mild conditions by direct *in situ* reaction on the HPTLC layer to produce "Pirkle" type chiral stationary phases (described more fully under chiral-bonded phases in this chapter).

However, it is the basic nature of the amino group that has led to a unique application area. With a pKa of ~9.5–11 for the amino group, the HPTLC layer can be used as a weak base anion-exchanger. In this mode satisfactory separations of nucleotides, mono and polysulfonic acids, purines, pyrimidines and phenols have all been achieved.[27–29] The eluents used are simple neutral mixtures of ethanol/aqueous salt solutions, which provide ideal conditions for ion-exchange to occur. The presence of sodium or lithium chloride helps to prevent secondary interactions and hence results in sharper, less diffuse spots or bands in the chromatogram. Examples of separations of this type are shown in Figures 14–16. Previously such separations were only achievable with PEI-cellulose (described later in this chapter). The amino-bonded high performance silica gel gives improved resolution.

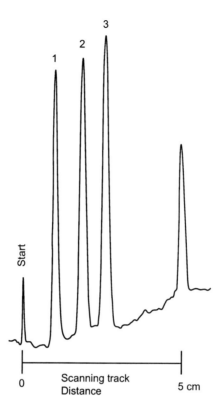

Separation of steroids on HPTLC pre-coated plates RP-18 F_{254s}

Sample substances:	1 methyltestosterone
	2 Reichstein »S«
	3 hydrocortisone
Applied quantity:	200 nl (10 mg/10 ml per substance)
Eluent:	acetone-water 60/40 (v/v)
Migration distance:	5 cm
Chamber:	normal chamber without chamber saturation
Evaluation:	UV 254 nm, TLC/HPTLC scanner (Camag)

Figure 12 *Separation of steroids on a silica gel RP18 HPTLC plate* (By permission of Merck)

Using the ion-exchange mechanism that predominates here, it is possible to influence the rate of migration of the analytes by using a salt solution as the eluent and varying the concentration. This can be done in a controlled manner to obtain the optimum separation conditions. As the ionic strength increases, anions are less retained and move to the solvent front. This is demonstrated in Figure 17 with adenosine and some oligoadenylic acids. As expected, the uncharged adenosine is largely unaffected by the ion-exchange mechanism, and shows little change in migration rate value as the lithium chloride concentration increases. Of course,

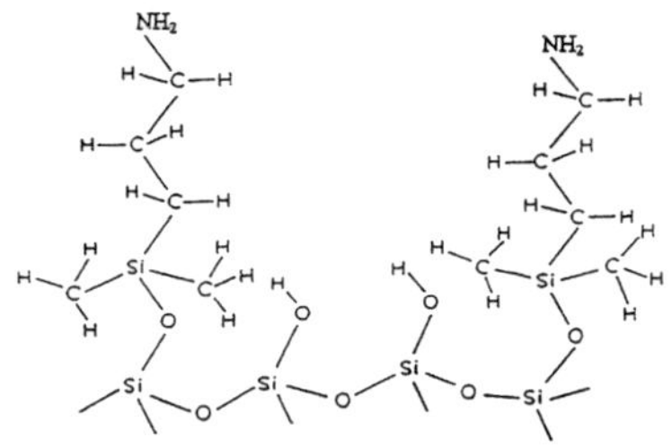

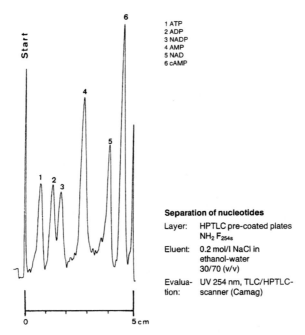

Figure 13   *Chemically bonded aminopropyl groups bonded to the silica gel surface*

changes in pH will also influence the mobility of analytes that are separated by ion-exchange.

Another unique feature of amino-bonded layers is their ability to provide reagent-free detection of certain chemical substances. The process is basically thermochemical and consists of heating the chromatographic layer after development to temperatures of at least 105 °C and sometimes as high as 220 °C. On exposure to short or long wavelength UV light after heating, the derivatives

Start

6

1 ATP
2 ADP
3 NADP
4 AMP
5 NAD
6 cAMP

Separation of nucleotides

Layer: HPTLC pre-coated plates
              $NH_2\,F_{254s}$

Eluent: 0.2 mol/l NaCl in
              ethanol-water
              30/70 (v/v)

Evalua-   UV 254 nm, TLC/HPTLC-
tion: scanner (Camag)

0   5 cm

Figure 14   *Separation of nucleotides on NH_2 bonded layers (Separation occurs because of the variation in affinity for the PO_4^{3-} groups)*
Sorbent layer: HPTLC silica gel 60 $NH_2\,F_{254}\,10 \times 10$ cm glass plate
Mobile phase: 0.2 M sodium chloride in ethanol/water $(30 + 70\,v/v)$
Detection: UV 254 nm, TLC/HPTLC scanner III (CAMAG)
(By permission of Merck)

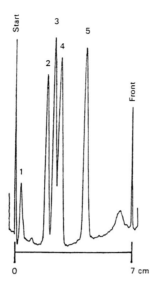

Figure 15 *Separation of purine derivatives on HPTLC silica gel 60 NH$_2$ plate*
Mobile phase: ethanol/water (80 + 20 v/v) saturated with sodium chloride
Sample applied: 300 nl of 0.1% w/v solutions
Detection: UV 254 nm (TLC/HPTLC scanner)
Peaks: 1, uric acid; 2, xanthine; 3, hypoxanthine; 4, guanine; 5, adenine
(By permission of Merck)

exhibit a strong fluorescence. This type of visualisation is particularly useful for quantitative analysis as the background on the chromatographic layer remains unaffected. The whole process is known as *thermochemical fluorescence activation* and is described in more detail in Chapter 6 (section 3.2). The technique has proved to be particularly suitable for the fluorimetric detection of carbohydrates, catecholamines, steroidal hormones, and fruit acids (see example in Figure 18 that illustrates the effect with a separation of carbohydrates).

Cyano-Bonded Phase. Cyano-bonded phases, prepared by the chemical bonding of a cyanopropyl group via a siloxane to silica gel, provide a further type of bonded phase for TLC. The first report of the use of cyano-bonded plates was for the separation of six polynuclear aromatics.[30] This was followed by a brief examination of the suitability of a cyano-layer for the separation of lipophilic dyes and PTH-amino acids.[31] More recently this phase has become commercially available as an HPTLC plate using silica gel 60 from Merck.[32] The surface coverage of this layer of cyano groups is 3.5 mmol m^{-2}. The bonding to the silica matrix is the same as that for the amino-bonded layers as shown in Figure 19.

The cyano phase fills a gap in the range of silica gel stationary phases with properties intermediate between reversed- and normal-phase materials. Hence with a choice of mobile phases either a reversed- or normal-phase separation can be carried out on a cyano-bonded TLC layer. This is demonstrated in Figure 20, for a

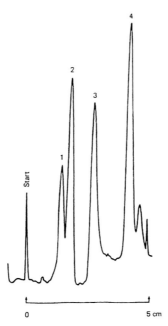

Figure 16   *Separation of naphthalene sulfonic acids on HPTLC NH₂ plates*
Mobile phase: ethanol/ammonia, pH 12 (60 + 40 v/v) + 0.18 mM sodium chloride
Sample applied: 200 nl of 0.5% w/v solutions
Detection: UV 254 nm (TLC/HPTLC scanner)
Peaks: 1, naphthalene-1,3,7-trisulfonic acid; 2, naphthalene-1,3,6-trisulfonic acid; 3, naphthalene-1,5-disulfonic acid; 4, naphthalene-1-sulfonic acid
(By permission of Merck)

series of steroids separated under both normal-phase conditions, with a high solvent strength eluent, and under reversed-phase conditions, with a low solvent strength mobile phase. For normal-phase, the analytes are less retained on the cyano phase than on silica gel 60, but more retained than for alkyl-bonded phases. The opposite effect is the case for reversed-phase conditions. Of course, the elution order is usually reversed when changing between normal- and reversed-phase modes.

Cyano plates have already been used for a wide variety of applications: benzodiazepine derivatives, pesticides, plasticisers, tetracycline antibiotics, phenols, some estrogens, gallic acid esters, alkaloids and sorbic acid.[32,33] Two further examples of the separations possible with cyano plates are given in Figures 21 and 22. In Figure 21 three steroids are base-line resolved under normal-phase conditions over a solvent front migration distance of 7 cm. By comparison Figure 22 shows a reversed-phase separation of eight alkaloids over the same migration distance. The addition of 0.1 M ammonium bromide to the solvent when reversed-phase conditions are used, improves the peak shape.

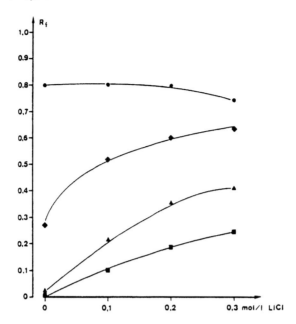

Figure 17 *Dependence of the R_f values of adenosine and oligomers on the lithium chloride concentration in the mobile phase on HPTLC NH_2 layers*
Mobile phase: methanol/water (90 + 10 v/v) + lithium chloride
Sample applied: 200 nl of 0.1% w/v solutions
Detection: UV 254 nm
Compounds: ● adenosine; ◆ ApA; ▲ ApApA; ■ ApApApA
(Reprinted from W. Jost, and H. E. Hauck, *Advances in Chromatography*, 1987, p 157 by courtesy of Marcel Dekker Inc.)

The dual personality of normal- and reversed-phase for the cyano-bonded phase allows unique two-dimensional separations to be achieved. The sample is applied as a spot near one corner of the plate. The layer is developed linearly in one dimension with a typical normal-phase solvent. Migration is allowed to continue until the solvent front reaches almost to the top of the plate. After appropriate drying, the layer is then developed in the second dimension (at 90° to the first direction) using a reversed-phase eluent. Migration of the solvent front is allowed to continue as before. The separation results show a "fingerprint" pattern of the components of the sample, unique to the solvent systems used. Figure 23 shows a separation of bile acid metabolites and cholesterol, a difficult resolution to achieve in one-dimensional TLC. Two-dimensional separations are easily performed in TLC and are particularly useful for samples containing many components that are not easily resolved by other techniques.

Diol-Bonded Phase. The polarity of the diol-bonded phase is very similar to that of the cyano phase. Figure 24 illustrates this similar retention of neutral steroids under

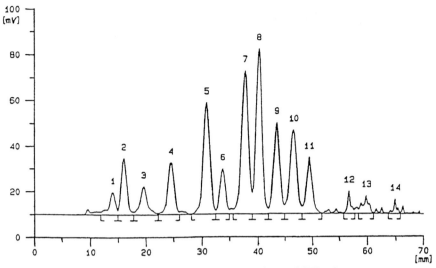

Figure 18   *Separation of carbohydrates on HPTLC silica gel NH₂ plate*
Mobile phase: AMD gradient based on acetonitrile–acetone–water. (15 step)
(AMD technique is described in Chapter 5)
Detection: UV 366 nm (Plate heated at 150 °C for 3–4 minutes)
Peaks: 1, maltohexose; 2, maltopentose; 3, maltotetrose; 4, maltotriose; 5,
maltose; 6, saccharose; 7, glucose; 8, fructose; 9, xylose; 10, rhamnose; 11,
deoxy-ribose
(Reprinted from CAMAG application note A-58.1 by courtesy of CAMAG,
Muttenz, Switzerland)

both normal- and reversed-phase conditions although the diol-bonded phase is used preferentially for normal-phase partition separations.

The plates are modified to the maximum degree possible with a vicinal diol alkyl ether group bonded via a siloxane to the silica gel surface by the usual silanisation procedure. The resulting structure of the diol group bonded to the silica gel matrix is shown in Figure 25.

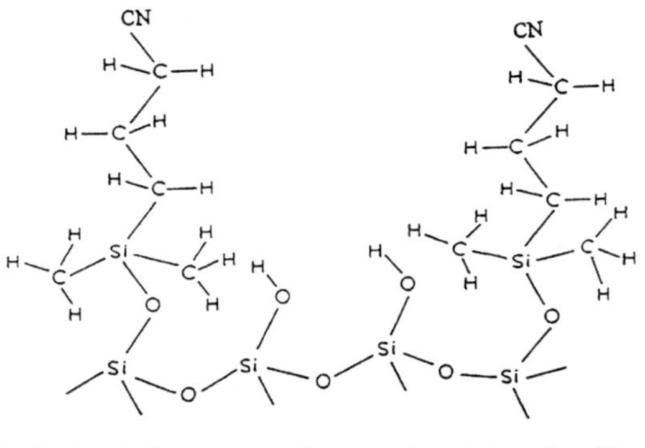

Figure 19   *Chemically bonded cyanopropyl groups bonded to the silica gel surface*

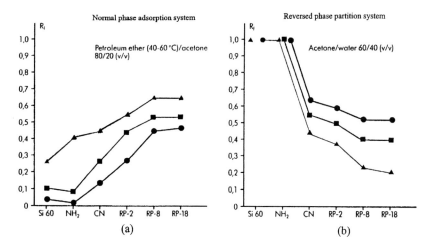

(a)　　　　　　　　　　　　　(b)

Figure 20 *Separations comparison of silica gel CN HPTLC plate with other bonded layers under,* (a) *normal-phase adsorption,* (b) *reversed-phase partition*
Mobile phase: (a) *petroleum ether (40–60 °C)/acetone (80:20 v/v)*
(b) *acetone/water (60: 40 v/v)*
Compounds: ● *hydrocortisone;* ■ *Reichstein's substance;* ▲ *methyltestosterone*
(Reprinted from Proceedings of the 2nd International Symposium on Instrumental HPTLC – Würzburg 1985, 1985, p86, by courtesy of Prof. Dr R. E. Kaiser, Institut für Chromatographie)

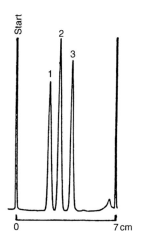

Figure 21 *Separation of steroid hormones on silica gel 60 HPTLC CN plates. Normal-phase separation*
Mobile phase: petroleum ether (40–60 °C)/ethanol (80 + 20 v/v)
Sample applied: 200 nl (20 mg/10 ml)
Detection: 5% perchloric acid in methanol followed by heating at 110 °C for 5 minutes. UV 366 nm, CAMAG TLC/HPTLC scanner
Peaks: 1, 1,4-androstene-3,17-dione; 2, 2-progesterone 3, 3-pregnenolone
(By permission of Merck)

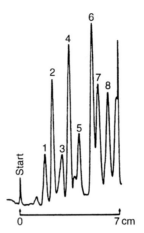

Figure 22   *Separation of alkaloids on silica gel 60 HPTLC CN plates. Reversed-phase*
separation
Mobile phase: methanol/propan-1-ol/25% ammonia soln./water (30 + 20 + 1 +
50 v/v) + 0.1 mol l^{-1} ammonium bromide
Sample applied: 300 nl
Detection: UV 254 nm, CAMAG TLC/HPTLC scanner
Peaks: 1, emetine chloride (40 mg/10 ml); 2, quinine (10 mg); 3, atropine
(300 mg); 4, brucine (10 mg); 5, scopolamine (300 mg); 6, caffeine (30 mg);
7, nicotinamide (20 mg); 8, tropic acid (300 mg)
(By permission of Merck)

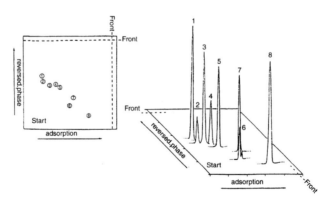

Figure 23   *2-D separation of cholesterol and bile acid metabolites on silica gel 60 HPTLC*
CN plate
Mobile phase: 1st dimension: acetone/water (50 + 50 v/v) 2nd dimension:
petroleum ether (40–60 °C)/acetone (70 + 30 v/v)
Detection: MnCl$_2$–sulfuric acid. Heat to 110 °C (5 min.). UV 366 nm
Compounds: 1, cholic acid; 2, dehydrocholic acid; 3, cholic acid methyl ester;
4, chenodesoxycholic acid; 5, desoxycholic acid; 6, 7-hydroxy-cholesterol;
7, lithocholic acid; 8, cholesterol
(By permission of Merck)

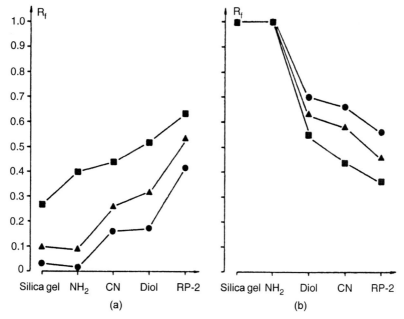

Figure 24 *Separations comparison of silica gel DIOL HPTLC plate with other bonded layers under,* (a) *normal-phase adsorption,* (b) *reversed-phase partition Mobile phase:* (a) *petroleum ether (40–60°C)/acetone (80 + 20 v/v)* (b) *acetone/water (60 + 40 v/v) Compounds:* ■ *hydrocortisone;* ▲ *Reichstein's substance;* ● *methyltestosterone* (By permission of Merck)

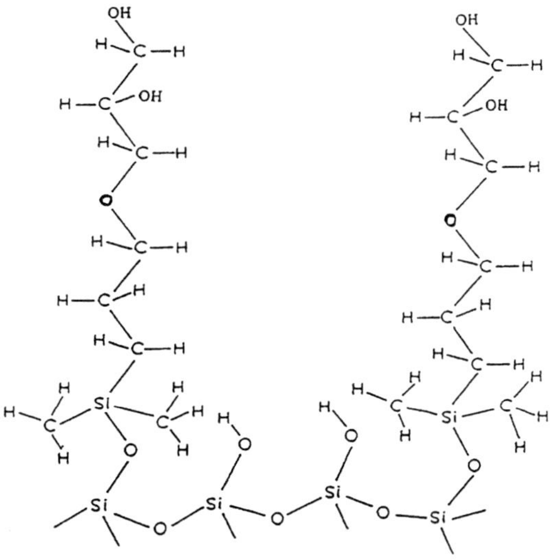

Figure 25 *Chemically bonded propyl–glycol groups bonded to the silica gel surface*

The layer is hydrophilic in nature and often behaves in a similar way to un-modified silica gel 60 in its chromatographic behaviour. However, as the main mechanism is normal-phase partition, the retention of sample components can be changed predictably. There are also two other main differences:

1. The hydroxyl groups of the diol phase take the form of a glycol. Silica gel 60 has as its "active" group a primary hydroxyl (silanol). As the hydroxyl function determines the retention, the nature of these may affect the chromatography.
2. The diol function is bonded via an alkyl ether spacer group. This also can influence the chromatographic behaviour. Thus as a general rule sample components migrate further on the diol phase as compared with silica gel 60 for the same developing solvent and solvent front migration distance.

Diol HPTLC plates have proved useful for a number of separations including digitalis glycosides, anabolic steroids, aromatic amines and particularly dihydroxy-benzoic acids.[34–36] Figures 26 and 27 demonstrate two of these. A near base-line separation of therapeutically useful digitalis glycosides can readily be obtained over a solvent migration distance of 8 cm. By comparison a more difficult separation of steroids is shown in Figure 27 with the estradiol only just resolved from the ethinylestradiol.

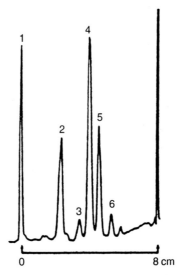

Figure 26   *Separation of digitalis glycosides on silica gel 60 HPTLC DIOL plates*
Mobile phase: ethyl acetate/ammonia soln. 25% (100 + 1 v/v)
Sample applied: 100 nl of 0.1% w/v soln
Detection: MnCl$_2$ – sulfuric acid. Heat at 110 °C for 10 min
UV 366 nm, CAMAG TLC/HPTLC scanner
Peaks: 1, lantoside C; 2, digoxin; 3, digitoxin; 4, digoxigenin; 5, α-acetyl-digoxin; 6, digitoxigenin
(By permission of Merck)

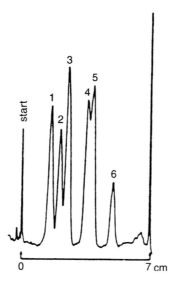

Figure 27 *Separation of anabolic steroids on silica gel 60 HPTLC DIOL plates*
Mobile phase: di-iso-propyl ether/acetic acid (100 + 1 v/v)
Sample applied: 300 nl of a 0.1% w/v soln
Detection: MnCl₂ – sulfuric acid. Heat at 110 °C for 5 min. UV 366 nm CAMAG
TLC/HPTLC scanner
Peaks: 1, 19-nortestosterone; 2, methoxyprogesterone; 3, progesterone; 4, 17β-
estradiol; 5, 17α-ethinylestradiol; 6, meso-hexestrol
(By permission of Merck)

Chiral-Bonded Phases. Because one enantiomer of a drug may be therapeutically active, whereas the other may be non-active, of a different activity or even toxic, the separation of optical isomers has become an increasingly important requirement in the pharmaceutical industry. There is also increasing interest in the agrochemical area where pesticides may vary in potency depending on the optical isomers present. At the present time there are many stationary phases that have been developed for HPLC and GC, which will resolve a large number of enantiomers. However, in planar chromatography by comparison few bonded phases are presently available commercially, although a number have been described in the literature.

In 1983, Wainer *et al.*[37] reported the successful preparation of an ionically-bonded chiral TLC plate using N-(3,5-dinitrobenzoyl)-R-(-)-phenylglycine, a phase originally devised by Pirkle *et al.*[38,39] This reagent was dissolved in tetrahydrofuran and allowed to migrate linearly through an amino-bonded TLC plate. The process was slow and complete derivatisation over the whole plate difficult due to the solvent migrating well ahead of the reaction zone. The bonded amino group is a primary amine and as such is reactive enough for an immediate reaction to occur. It is possible to prepare such plates both with an ionic bond or a more stable covalent bond by a dipping technique. The covalent bonding process is achieved

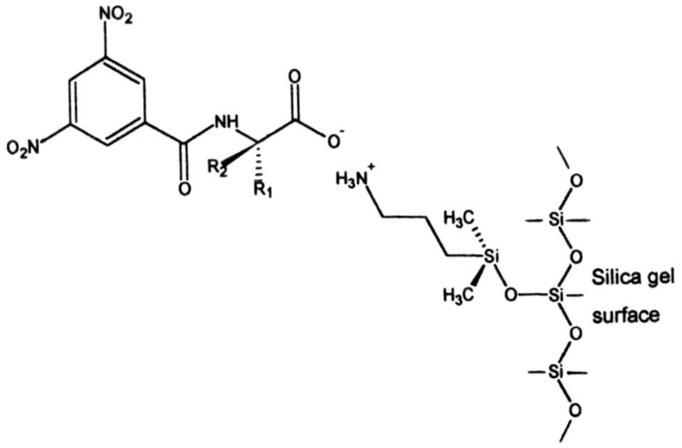

Figure 28   *Structure of Pirkle reagent bonded to aminopropyl silica gel. (R_1 is –H, and R_2 is –alkyl or –aryl)*

under a nitrogen atmosphere and in the presence of a catalyst. Both the R-α-phenylglycine and the L-leucine derivatives have been prepared. These plates are relatively stable if stored stacked under dry conditions in a desiccator. For most applications, ionic bonding is quite adequate for the range of chiral separations possible with this phase. The plates are prepared simply by dipping the amino-bonded HPTLC plates in a 0.3% solution of the chiral selector in dry tetrahydrofuran (THF). Excess reagent is removed with dry solvent. The resulting chiral phase has the structure shown in Figure 28.

   Chiral recognition of enantiomers on these "Pirkle" phases depends on a three-point interaction involving hydrogen-bonding, π-π interaction between aromatic or

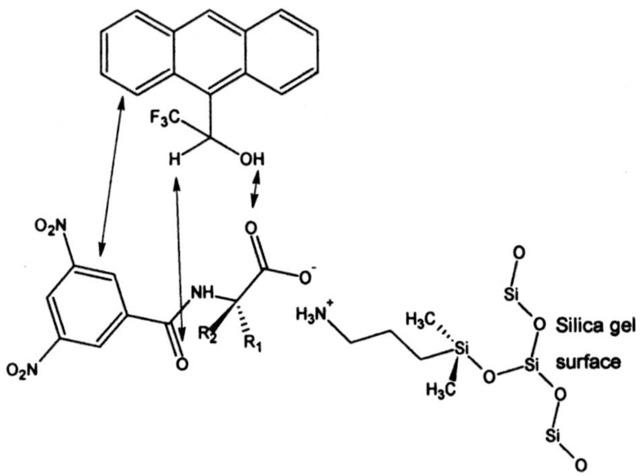

Figure 29   *Interactions of 2,2,2-trifluoro-1-(9-anthryl) ethanol with a Pirkle reagent-bonded layer (R_1 is –H, and R_2 is –alkyl or –aryl)*

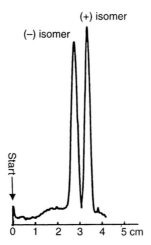

Figure 30 *Separation of (+) and (−) enantiomers of 2,2,2-trifluoro-1-(9-anthryl) ethanol using CSP1 amino-bonded HPTLC plate. Mobile phase: n-hexane/propan-2-ol (80 + 20 v/v). Scanned at 380 nm with a CAMAG TLC/HPTLC scanner*

unsaturated groups, in which one is a π-donor and the other a π-acceptor, and dipole stacking. N-(3,5-dinitrobenzoyl)-R-(-)-α-phenylglycine can provide sites for hydrogen-bonding at the >NH and >C=O functions and the dinitrophenyl group acts as a π-acceptor. If in the sample, groups that can hydrogen bond are present close to the chiral centre and an aromatic group is present which is a π-donor, then one enantiomer will be retained more strongly than the other due to the interactions being more favourable. Such three-point interactions usually result in well resolved enantiomers (see Figure 29). However, adequate separations can often be obtained with fewer interactions. Using this type of chiral-bonded TLC plate, separation of test compounds such as (±)2,2,2-trifluoro-1-(9-anthryl) ethanol (containing a single asymmetric carbon centre) and (±)2,2'-bi-2-naphthol (dissymmetry due to hindered rotation) into their respective enantiomers has been demonstrated.[40] (Enantiomer separation of the former is shown in Figure 30.)

Hexobarbital, some benzodiazepines and β-blocking drugs have also been separated on both the L-leucine and R-α-phenylglycine chiral phases.[40,41] However, the β-blockers required derivatisation with 1-isocyanatonaphthalene before separation. The major problem with this type of chiral stationary phase is its comparatively low sensitivity and limited application. In particular many detection reagents cannot be used due to reaction with the background amino-acid present on the silica gel surface producing a background colour. In addition the "Pirkle" chiral selector masks the inorganic fluorescent indicator (F_{254s}) incorporated into the amino-bonded HPTLC plates. A possible way of overcoming the problem involves bonding a section of the HPTLC plate with the Pirkle reagent, spotting the racemate at the lower edge of this area, and proceeding with normal linear migration with an appropriate solvent until the enantiomers separate, and pass through into the unreacted zone. The separated isomers are then detected in the normal way under UV light or with derivatisation reagents. In this way sensitivity is improved. The solvents for separations on these "Pirkle" phases consist almost entirely of n-hexane – propan-2-ol mixtures.

Brunner and Wainer have also prepared a naphthylethyl urea-bonded TLC plate following the method described by Oi *et al.*[42,43] R-(-)-1-(1-naphthyl) ethyl isocyanate was reacted with an HPLC type aminopropyl silica gel. The TLC bonded phase was simply prepared by dipping an aminopropyl TLC/HPTLC plate into a 1% solution of the naphthyl isocyanate in dichloromethane for five minutes. This phase does not suffer from the background interference problems in the UV associated with the previous Pirkle phases. The naphthyl moiety is π-basic (a π-donor) and hence it is particularly useful for the separation of enantiomers containing π-acidic groups, *e.g.* dinitrobenzoyl derivatives of amino-acids. A series of α-methylarylacetic acids, (ibuprofen, fenoprofen and naproxen) have also been separated after conversion to their respective 3,5-dinitroanilides.

The preparation of a β-cyclodextrin phase, bonded to silica, has also been described.[44,45] The chromatographic results that were obtained were dependent on the source of silica gel, coverage of the layer, and the type of binder used. Mobile phases used with this phase consisted of methanol or acetonitrile in water or buffers. The use of buffers improved resolution and efficiency. Some separations of enantiomers of dansyl amino-acids, alanine-β-naphthylamides and metallocenes have been reported on this phase. However, most chiral TLC work with cyclodextrins has been accomplished using them as mobile phase additives.

Currently the most successful chiral TLC layer (and the only one commercially available) is not a bonded phase, but a coated phase consisting of an amino-acid derivative of hydroxyproline in the presence of a copper salt. This forms a ligand-exchange phase when coated on to a RP18 bonded reversed-phase plate. The method of preparation is based on the work of Davankov *et al.*[46,47] who developed this technique with HPLC columns. It involves dipping an RP18 bonded TLC plate into a copper(II) acetate solution (0.25% w/v) followed by a solution of the chiral selector (0.8%), usually N-(2'-hydroxydodecyl)-4-hydroxyproline (Figure 31). After air drying the plate is ready for use.

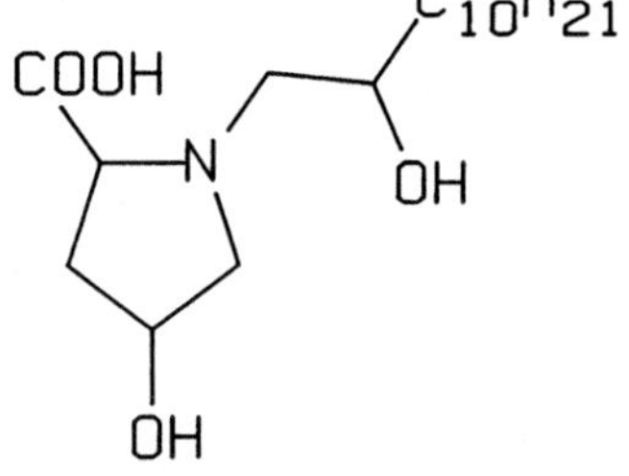

Figure 31 *Structure of chiral selector, (2S, 4R, 2'RS)-N-(2'hydroxydodecyl)-4-hydroxy-proline used in the preparation of ligand-exchange plates*

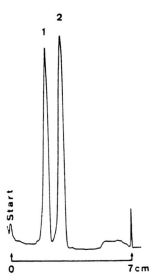

Figure 32 *Separation of dipeptides on CHIR HPTLC plate*
Mobile phase: methanol–propan-1-ol–water (50 + 10 + 40 v/v)
Detection: ninhydrin spray (0.5% w/v in butan-1-ol). Heat at 120°C for
10 minutes. Visible scan at 420 nm (CAMAG TLC/HPTLC Scanner)
Peaks: 1, D-Leu - L-Leu; 2, L-Leu - D-Leu
(By permission of Merck)

The mechanism of enantiomer separation involves the formation of diastereoisomeric complexes with the metal ion. The stability of these transient complexes differs depending upon the individual optical isomer and hence retention on the stationary phase will vary and the enantiomers can be resolved. This ligand-exchange plate is available from Macherey-Nagel[b] as "Chiralplates" and Merck as "CHIR". The Chiralplate is a TLC glass plate, whilst the CHIR is either an HPTLC or a TLC glass plate. Both types of ligand-exchange plate have proved suitable for the resolution of most amino-acids, halogenated, N-alkyl and hydroxy amino-acids, thiaxolidine derivatives, dipeptides and catecholamines.[48–51] As an example of the excellent resolution often obtained, Figure 32 shows the separation of two optical isomers of a peptide with a simple solvent mixture. Quantification has also proved successful, with both the TLC and HPTLC layers and, for example it has been possible to measure nanogram amounts of L-tryptophan in the presence of microgram quantities of D-tryptophan. Figure 33 shows the chromatogram, whilst the calibration curve for determining an unknown concentration of one enantiomer is shown in Figure 34.

---

[b] Macherey-Nagel, Duren, Germany.

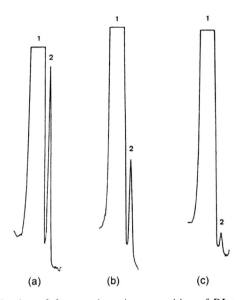

Figure 33 *Determination of the enantiomeric composition of DL-tryptophan at extreme ratios of the antipodes, a: 1:100, b: 1:200, c: 1:1000*
Mobile phase: methanol–water–acetonitrile (50 + 50 + 30 v/v)
Peaks: 1, D-tryptophan, (a) 10 μg, (b) 10 μg, (c) 10 μg; 2, L-Tryptophan, (a) 100 ng, (b) 50 ng, (c) 10 ng
(M. Mack, H. E. Hauck, and H. Herbert, *J. Planar Chromatography,* 1988, **1,** 307, by permission of the publisher)

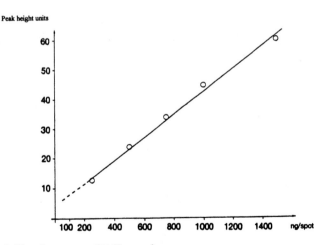

Figure 34 *Calibration curve of D-Tryptophan*
Sample applied: 75 ng
Concentration: 100, 250, 500, 750, 1000, and 1500 ng
Detection: ninhydrin spray (0.5% w/v in butan-1-ol). Heat at 120 °C for 10 minutes. Scanned at 520 nm (CAMAG TLC/HPTLC Scanner)
(M. Mack, H. E. Hauck, and H. Herbert, *J. Planar Chromatography,* 1988, **1,** 308, by permission of the publisher)

1.3 Non-silica Sorbents

1.3.1 Cellulose

Cellulose, a product of natural origin, has a polymeric structure consisting of glucopyranose units joined together by oxygen bridges. As shown in Figure 35, a profusion of hydroxyl groups are present which are readily available for hydrogen-bonding. Adsorbed water or alcohols can be retained by this interaction, making cellulose an ideal phase for the separation of hydrophilic substances such as amino-acids, carbohydrates, inorganic ions and nucleic acid derivatives.[52-59]

Two types of cellulose are used in planar chromatography. One is native fibres with a typical polymerisation of between 400–500 glucopyranose units, (used for paper chromatography and in some TLC layers). The other is a microcrystalline form commonly called 'Avicel', a fine powder used widely in both TLC and HPTLC, and prepared by a hydrolysis technique. It has a degree of polymerisation of 40–200 glucopyranose units. The cellulose is obtained from a number of raw materials, including wood and cotton. However, the former does require more refining and has a lower cellulose content.

For the preparation of TLC/HPTLC plates, a similar slurry technique to that needed for silica gel is employed. However, unlike silica gel, binders are unnecessary. The chromatographic results obtained with either the fibrous or microcrystalline types can be different. However, whatever the type, the resolution of samples is generally not as sharply defined as that obtained with silica gel. Spots and bands are more diffuse and separation times are usually longer. Pre-coated plates are available

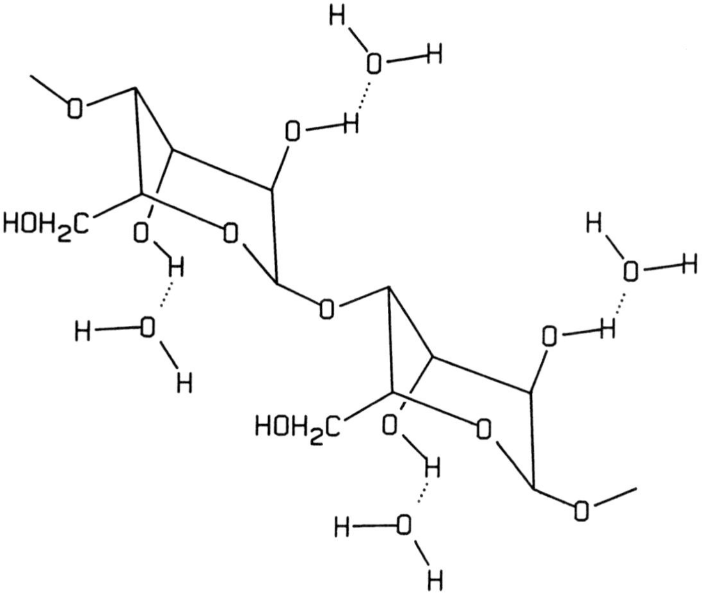

Figure 35 *Structure of cellulose illustrating the hydrogen-bonding effect with water*

from most of the TLC plate suppliers, but few supply a high-performance layer. The diffusion of chromatographic zones is greatly reduced with HPTLC cellulose, but one must remember to adhere strictly to the application of small quantities (~100 ng) of sample with final spot diameters of approximately 1 mm. Commercial pre-coated cellulose plates are usually made as thinner layers than for silica gel, nominally 0.1 mm thick.

In recent years cellulose has declined in popularity as methods have switched to those based on silica gel or to other techniques. However, many separations have been based on cellulose and it is still widely in use for amino-acid separations, particularly in hospital clinical laboratories where abnormal increases in certain amino-acids in blood or urine samples can indicate the presence of a number of diseases (an increase in phenylalanine is indicative of phenylketourea). Two-dimensional methods are used with two developments in each dimension with intermediate drying. A typical amino-acid "fingerprint" pattern obtained is shown in Figure 36. The amino-acids are detected with ninhydrin spray after development or the ninhydrin can be incorporated into the final development step as part of the solvent mixture. This latter procedure avoids the problem of using potentially hazardous spray reagents and shortens the analysis time. However, it is important in this instance to ensure that the final developing solvent does not contain ammonia, amines, or amides as these will usually form coloured chromophores with the ninhydrin. The resulting chromatogram will be poor due to the high background colour.

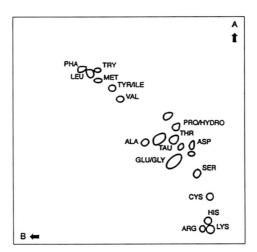

Figure 36   *2-dimensional separation of amino-acids on cellulose coated TLC aluminium sheets*
   Mobile phase: solvent A: butan-1-ol–propionic acid–water (40 + 10 + 10) [1st dimension]
   solvent B: pentan-1-ol–butanone–pyridine–water (20 + 20 + 20 + 15) [2nd dimension]
   Detection: ninhydrin spray (0.5% w/v in butan-1-ol). Heat at 120 °C for 10 minutes

1.3.2 Cellulose Bonded Phases

PEI Cellulose. PEI cellulose is a polyethyleneimine modified cellulose which acts as a strongly basic anion-exchanger. It has had fairly specific uses including the analysis of nucleotides, nucleosides and nucleo-bases, vanadylmandelic acid (VMA), and sugar phosphates.[60] For these applications PEI cellulose was the sorbent of choice for many years. Most of these separations are now achievable on amino-bonded silica gel layers with improved resolution. The PEI cellulose TLC layers require storage at 0–4 °C to reduce deterioration. As the plates become old, the layers take on a pale brown coloration, and should be discarded.

Acetylated Cellulose. Acetylated cellulose, particularly triacetyl-cellulose, is prepared by chemical reaction of the hydroxyl groups on cellulose to produce a layer with reversed-phase characteristics. Its major use for many years has been for the separation of polyaromatic hydrocarbons, an important area of interest. However, in many cases it has been superseded by bonded reversed-phase silica gels.

In more recent times the use of acetylated cellulose as a chiral layer for the separation of optical isomers has been investigated.[61] The resolution of enantiomers is very dependent on the cellulose structure and the acetyl content of the cellulose triacetate. The best results are obtained with a layer of microcrystalline cellulose triacetate (particle size of 10 μm) with a silica gel 60 binder. Although the sodium salt of carboxymethylcellulose has also been used as binder, silica gel enables the use of aqueous based eluents. As with column separations of enantiomers, mixtures of ethanol or propan-2-ol (70–80%) with water (20–30%) serve well as mobile phases. Resolution varies according to organic solvent concentration in the eluent. It has also been observed that temperature has a noticeable effect on the quality of separation. Generally as temperature increases from 25 to 40 °C, resolution of enantiomers decreases.[61]

Racemates of a number of organic species have been separated on TLC layers of microcrystalline cellulose triacetate. These include specific compounds such as benzoin, benzoin methyl ether, flurbiprofen, 1-(2-naphthyl) ethanol, aminoglutethimide, 1,1'-binaphthyl-2,2'-diamine, N-[1-(naphthyl) ethyl] phthalamic acid, and a few derivatised amino-acids.[61] However, to date the use of this chromatographic layer has not developed commercially probably due to the long development times that are required (about 2.5 hours) for sufficient migration of analytes.

Carboxymethyl (CM) and Diethylaminoethyl (DEAE) Cellulose. Carboxymethyl and diethylaminoethyl celluloses are prepared as ion-exchange media, the former is weakly acidic and the latter is strongly basic. The exchange capacities are often close to those for ion-exchange resins, but their behaviour is often quite different. This is due to the hydrophilic nature of the base cellulose compared with the hydrophobic nature of the base polymer in the resin material.

1.3.3 Aluminium Oxide

Aluminium oxide or alumina, like silica gel, is a synthetic sorbent. It is manufactured in three pH ranges; acidic, basic, and neutral for different types of samples.

Thus under aqueous conditions acidic compounds like phenols, sulphonic, carbox-ylic, and amino acids are separated on the acidic alumina, whilst basic compounds; amines, imines, and basic dyes, are separated on basic alumina. Neutral compounds, such as aldehydes, ketones and lactones are chromatographed on neutral alumina. Of the three types, basic alumina is the most widely used. In non-aqueous eluents, aromatic hydrocarbons, carotenoids, porphorins, alkaloids, and steroids can be adsorbed. As with silica gel, alumina will also vary in activity according to water content. The activity levels have been graded according to Brockmann and Schodder.[62] Table 5 shows the Brockmann grading according to water content of alumina.

Some manufacturers offer a type T and E in their alumina TLC plate range. Usually there is a difference in pore size, but it should be noted that type T is ignited at a higher temperature. The Brockmann grade for type E resembles activity I whereas type T resembles activity II. Alumina is more chemically reactive than silica gel and this can lead to problems with some samples. Reactions can occur in the sorbent layer causing loss of analytes during chromatography. Like cellulose, alumina is now also declining in popularity, and one would need to consult the older bibliographies for applications.

1.3.4 Kieselguhr

Kieselguhr is a natural diatomaceous earth, composed of the skeletal remains of microscopic marine organisms deposited in times past. Although principally silicon dioxide, it also contains varying amounts of other oxides of aluminium, iron, titanium, magnesium, sodium, potassium and calcium as oxides, hydroxides, and carbonates (approximately 10% in all).[63] It is widely used as a filter aid due to its high porosity (average pore diameter is quite variable, typically $65\,000\,Å$). Kieselguhr is used in conjunction with 15% of a calcium sulphate binder to produce TLC plates. The variability of pore size and surface area limits the use of kieselguhr for high quality, precision TLC. It has been used in the past for the separation of polar compounds by a partition mechanism. Commercial pre-coated plates with abrasion resistant organic binders have been available for many years, although their usage has diminished in recent times.

Table 5 *Brockmann and Schodder activity grading for aluminium oxide for chromatography*[62]

Relative humidity	Aluminium oxide 60	Aluminium oxide 90	Aluminium oxide 150
0%	I	I–II	II
20%	II–III	III	III–IV
40%	III–IV	III–IV	IV
60%	IV–V	IV	IV
80%	V	V	IV–V

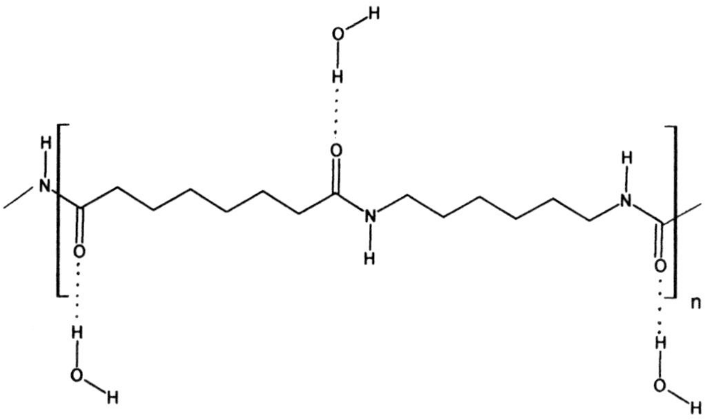

Figure 37 *Hydrogen-bonding of water with polyamide (nylon 66). Compounds that will form stronger hydrogen-bonds will require a stronger elutive solvent to cause migration*

1.3.5 Polyamide

The polyamide phases are produced from polycaprolactam (nylon 6), polyhexamethyldiaminoadipate (nylon 66), or polyaminoundecanoic acid (nylon 11). The chromatographic separation on polyamide depends on the hydrogen-bonding capabilities of its amide and carbonyl groups (Figure 37). The bond strength generated depends upon the number and position of any phenolic, hydroxyl or carboxyl groups present in the sample components. The relative retention of the analytes depends on the eluting solvent being capable of dissociating these bonds. As the solvent migrates through the sorbent, the analytes separate according to their ease of displacement.

Mixtures of phenols, indoles, steroids, nucleic acid bases, nucleosides, dinitrosulfonyl (DNS), dinitrophenyl (DNP), and dimethylaminoazobenzene isothiocyanate (DABITC) derivatised amino-acids, and aromatic nitro compounds have all been resolved on polyamide.[64–67] A range of pre-coated sheets with aluminium or plastic backing are commercially available, including one quite unique 15 cm square plastic sheet which is coated on both sides with polyamide 6. With this plate, samples containing, for example, amino-acid derivatives are applied to one side whilst the standards are put on the other. After chromatography the known amino-acids can be picked out immediately. This novel approach has been successfully applied to PTH, dansyl, and DNP amino-acids.

1.3.6 Miscellaneous Stationary Phases

Other less commonly used phases include magnesium silicate, chitin and Sephadex™. Magnesium silicate is a white, hard powder often known under the name of Florisil™. The manufacturer, Floridin, (Pittsburgh, USA) gives the surface area of the TLC grade as 298 m^2 g^{-1} and the pH as 8.5. It has been reported as suitable for the separation of carbohydrates and derivatives.[68]

Chitin is a polysaccharide composed mainly of 2-acetamide-2-deoxy-D-glucan molecules linked via oxygen bridges in a similar type of structure to cellulose but with a basic nature. Typical specific surface area is low, only 6 m^2 g^{-1}. Chitin has been used principally for amino-acid separations, but it has also been applied to inorganic ions, nucleic acids, phenols and dyes.[69]

Sephadex is a trade name of Pharmacia Fine Chemicals for a range of gel filtration materials. They are modified dextrin gels, hydrophilic and neutral in nature. They are rarely used in TLC as layers are difficult to prepare, and require pre-swelling for many hours before use. Sephadex has been used for the separation of peptides and nucleic acids.[70]

1.3.7 Mixed Stationary Phases

Mixed phases, for example silica gel/kieselguhr, silica gel/alumina and cellulose/silica gel are sometimes used for specific applications. However, they almost always require the preparation of a special layer with a specific ratio of components. Few commercial pre-coated plates are available. Silica gel/kieselguhr has been used for inorganic ions, herbicides and some steroids.[71–73] Cellulose/silica gel has found application in the separation of food preservatives, and antibiotics.[74,75] Silica gel/alumina has been little used in the last twenty years.

1.3.8 Dual Phases

Dual phase plates are a fairly recent innovation involving two distinct stationary phases on one TLC plate. Usually the two phases are normal and reversed-phase as in the case of the Whatman[c] Multi KCS5 and SC5 dual phase plates, but obviously other combinations are possible. It is important that there is a sharp, straight, level interface between the phases. Plates are used in two dimensions, in one direction as a normal-phase separation on silica gel, then after drying, in the second direction as a reversed-phase separation on silica gel RP18. This results in a two-dimensional "map" or "fingerprint" of the sample components. Two-dimensional TLC allows for the resolution of a larger number of components than normal linear development. Examples of the use of this dual phase plate include the separation of sulfonamides and bile acids.[76]

2 Preparation of TLC Sheets and Plates

2.1 "Home Made" TLC Plates

Before the advent of commercial pre-coated TLC plates, layers of silica gel and of other sorbents (*e.g.* cellulose or aluminium oxide) had to be prepared in advance of separation procedures. A typical method was to thoroughly mix 30 g of sorbent with 60 ml of water by shaking in a glass flask. The slurry produced was transferred to a spreader and applied evenly over the plate surface in one spreading operation.

[c] Whatman, Maidstone, Kent, UK.

The applied slurry was allowed to set and dry for about 30 minutes. Final activation was completed in an oven for 30 minutes at a maximum temperature of 105 °C. To improve the binding of the silica gel to the inert backing a gypsum binder was added; Stahl recommended 13% of calcium sulphate. Although the layer was soft and easily damaged, it remained stuck to the glass backing when the mobile phase migrated across the plate. This silica gel plate was given the designation "silica gel G". Organic binders have also been used including carboxymethylcellulose, starch (1–2%), and polyvinyl alcohol (1–5%). Although these give a stronger binding for the most part at a lower concentration of binder, they often suffer from solubilisation in aqueous based solvents and charring after treatment with strong sulphuric acid solutions and heating.

2.2 Pre-coated TLC/HPTLC Sheets and Plates

Commercially manufactured pre-coated TLC and HPTLC plates use organic polymeric binders at a concentration of about 1–2%. These binders are much more resistant to chemical elution or attack, ensure a smooth abrasive resistant surface, and effectively bind smaller particle size silica gels to an appropriate backing such as glass, aluminium or plastic. Quite strong polar solvents and detection reagents can be used without appreciable affect. The added advantage is that the plates provide a layer of highly active silica gel that is not deactivated by appreciable concentrations of inorganic binders. There is always a noticeable improvement in the quality of separation between such plates and those consisting of silica gel G. Most TLC and HPTLC silica gel plates are manu-factured with a uniform surface and a layer thickness of 0.20–0.25 mm (the major exception is cellulose with a layer thickness of 0.1 mm). These commercially manufactured plates give far more reproducible results compared with those that are "home-made". This is a major benefit in analytical work where repeatable quantitative results are essential. Particle size, pore diameter, smoothness and thickness of surface, and abrasive resistant qualities of a layer need to be controlled within fine tolerances by the manufacturer to ensure such high quality and reproducible results.

3 Cutting TLC/HPTLC Sheets and Plates

Although TLC plates can easily be cut with a glass cutter, and plastic and aluminium sheets can be cut with a pair of scissors, care does need to taken to avoid damage to the chromatographic layer. A clean pair of rubber or plastic gloves is essential when cutting glass plates to prevent contamination of the layer and any damage to the hands. Cutting aluminium and plastic sheets needs to be done in such a way that the minimum amount of damage to the surrounding layer occurs. A large pair of scissors is preferable with as few scissor cuts as possible. Always angle the scissors at approximately 30° from the vertical to one side as shown in Figure 38.

Poor cutting will result in a capillary crack between the chromatographic layer and the foil. This will cause the mobile phase to migrate more rapidly at the edge

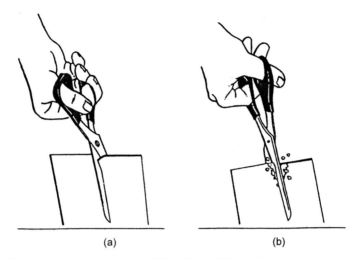

(a)   (b)

Figure 38   *Cutting aluminium-backed TLC sheets. The angle of cut is important to limit damage to the layer. A correct cut to the left tilted 30° from the vertical is shown in example* (a), *whereas a cut to the right in example* (b) *causes a cracking of the layer in the vicinity of the cut*

compared with the centre of the layer, and the mobile phase to migrate from the edge of the layer to the centre. This will cause spots or bands near the edge of the chromatogram to become deformed and chromatographic tracks to become distorted.

4 Humidity Effects with TLC Plates

Relative humidity is defined as the amount of water vapour saturation in air at a given temperature. It is usually expressed as a percentage and in most instances will vary from 40–60% for comfortable conditions. In the laboratory, TLC sorbents will adjust to an equilibrium water vapour concentration depending on the relative humidity. The adsorption is reversible so enforced drying can be used to lower the relative humidity of the sorbent layer. Often TLC plates and sheets are heated at 120 °C for at least half an hour to try to achieve a more constant activity. Really there is little point to this procedure as once the TLC plate is exposed to ambient conditions it takes only a few minutes at the most to readjust its water content to the surrounding environment. For the same reason it is usually a meaningless exercise to try to rejuvenate old TLC layers by heating.

Many manufacturers pack TLC plates at a standard relative humidity (*e.g.* 40% RH) and once exposed to the atmosphere the percentage water uptake will normally rise to ambient. Under normal conditions this will cause only minor variation in the activity but high relative humidity in excess of 60% causes a rapid change. In these instances heating the plates will drive off this excess moisture. To maintain good reproducibility it is therefore important to keep packs of plates well closed when

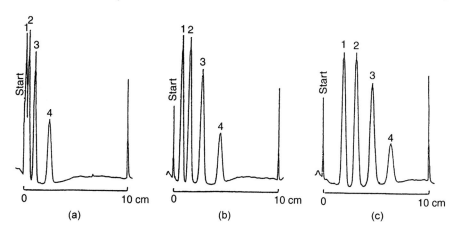

Figure 39 *Effect of relative humidity (RH) on the separation of oligophenylenes on silica gel 60 HPTLC plates*
Mobile phase: cyclohexane
Humidity: (a) 20% RH; (b) 50% RH; (c) 80% RH
Peaks: 1, m-quinquephenyl; 2, m-quaterphenyl; 3, m-terphenyl; 4, biphenyl
(E. Hahn-Deinstrop, *J. Planar Chromatography*, 1993, **6**, 317, with permission of the publisher)

not in use, and to maintain the laboratory environment within reasonable limits of relative humidity. Closing up packs of plates when not in use is also a good practice to avoid adsorption of any other vapours that may be present in the immediate atmosphere. Acidic and alkaline vapours will cause a change in the activity of the sorbent.

Special chambers have been developed which alleviate the problem of humidity to a large extent (twin-trough, horizontal, and automated development chambers). These chambers incorporate either sulfuric acid solutions or saturated salt solutions to control the relative humidity.

Changes in relative humidity can affect a number of important factors in TLC, *e.g.* the R_f value (retention factor), selectivity, solvent front migration rate and the position of multiple fronts. Where normal-phase or adsorption mechanisms prevail, increase in humidity causes less retention of analytes during development and faster elution. Selectivity is affected and a change in the sample component migration rates results. Figure 39 illustrates this well with a separation of aromatic hydrocarbons. Sometimes even a change in order of separated analytes can occur.

5 Pre-washing TLC Plates

Sometimes pre-washing of the TLC/HPTLC layer is necessary to remove impurities usually originating from the binder. Normally this is only a problem if the detection reagent is sensitive to the impurities or fluorescence quenching is being employed as the detection technique. The use of polar mobile phases causes a concentration of these impurities at the solvent front. As the resolution of components of the sample would be expected to be poor near the solvent front, the concentration of these impurities is not usually a problem. Non-polar mobile phases do not cause any migration of the impurities and hence there is little background interference when scanning. The best way to pre-wash a plate is by blank development in a TLC tank. Either the mobile phase for the separation can be used, or methanol, or mixtures of methanol and chloroform. Acids and bases are best avoided. The plate should be marked to show the direction of blank development. When the plate is then later used for chromatography of the sample, the same direction of development can and should be employed.

6 Use of Phosphorescent/Fluorescent Indicators

To aid absorbance many commercial pre-coated TLC layers contain an inorganic phosphorescent indicator, shown on the label by the designation F_{254} or UV_{254}. Most indicators exhibit a bright green, yellow or blue phosphorescence when excited by UV light at 254 nm. Indicators that have been used are uranyl acetate (yellow-green), manganese zinc silicate, zinc cadmium sulphide, zinc silicate (green), alkaline earth metal tungstates, and tin strontium phosphate (blue). Detection by absorbance in these cases relies on the phosphorescence being quenched by the sample components. In most cases pink or violet spots/bands are observed for quenching on the green phosphorescent background, whereas grey-black spots/bands are usually seen on the blue background. In most applications the inorganic indicators are quite stable with little or no elution from developing solvents, and they remain unaffected by most dyeing reagents and the temperatures used to effect reactions. Organic fluorescent indicators can also be used. These are coded F_{366} or UV_{366} by the manufacturer. Substances used include optical brighteners, hydroxypyrene sulfonates, fluorescein and rhodamine dyes. As indicated by the coding used, these organic indicators fluoresce in long wave UV light (at 366 nm). Some pre-coated plates are manufactured with both types of indicator present to give possible quenching by sample components at both wavelengths. Sometimes the designation F_{254s} is used in the description of some TLC and HPTLC plates. The "s" simply indicates that the fluorescent indicator is acid resistant, particularly useful where acid vapours may be used to adjust pH or cause further reaction of the analytes to render them able to quench fluorescence. Many analytes, however, either absorb insufficiently or not at all by these techniques. In these instances suitable detection reagents are used to give coloured spots/bands *in situ.*, (see Chapter 6).

7 Channelled TLC Layers

Some suppliers have made available channelled silica gel plates for special applications. Channelled TLC plates consist of silica gel tracks, usually 1 cm wide from the top to the bottom of the plate, with spaces of 1–2 mm between tracks. Of course, it is possible to make such channels oneself with a sharp metal point (*e.g.* a bradawl is quite adequate for the purpose) and a ruler. The width of such channels is therefore decided by the user. Commercially manufactured plate scrapers are also available that enable six channels to be cut on a 10×10 cm glass-backed layer. These units are to be recommended as they give far better precision than the purely manual method. The major advantage of such channels is to prevent the cross contamination of samples that may occur where sample spots are applied close together or where excessive diffusion occurs during development, causing some merging of components. A further advantage is when analyte recovery from the TLC layer is employed. Sections of a channel are much more easily removed without contamination from other separated components on the layer and without damaging other tracks.

8 Concentration Zone TLC/HPTLC Plates

The concentration zone concept offers a number of benefits for certain types of analysis. The benefits were first described by Abbott and Thomson in 1965[77] and elaborated on further in 1969 by Musgrave.[78] The TLC/HPTLC concentration zone plate consists of two different layer sections with a sharply defined borderline. There is no gap in between. The lower zone is used for sample application and covers the full width of the plate up to 25 mm. This lower zone is composed either of kieselguhr or synthetic porous silicon dioxide of medium pore volume, but very high pore diameter (50 000 Å) and extremely small internal surface area (less than $1 \mathrm{~m}^2 \mathrm{~g}^{-1}$). Kieselguhr being a natural material, mainly composed of silica, and approximately 10% of other metal oxides' can cause variations in chromatographic behaviour. The upper section of the plate is coated with normal silica gel 60 sorbent. Samples are applied as spots or bands to the concentration zone. Usually these are more dilute (typically 5–10 times) than what would be applied to a normal silica gel 60 TLC layer. On elution the analytes of interest migrate at or near the solvent front. On reaching the interface they form a concentrated band. This continues to migrate into the silica gel 60 layer and development continues normally. Although there are obvious advantages to the use of such concentration zone plates such as eliminating the need for precise positioning of sample application and sample volumes, the more detailed advantages and disadvantages of this layer will be discussed in later chapters.

9 HPTLC Pre-coated Plates

HPTLC uses the same type of silica gel 60 layers, as in traditional TLC, with a thickness of 0.20–0.25 mm. However, the particle size is much smaller, typically ranging from 4–8 μm, with an optimum of 5–6 μm. Recently a new spherical HPTLC silica gel 60 of optimum particle size 4 μm has become available (see comparison in Figure 40). The smaller particle gives a somewhat softer, smoother layer. However, commercial pre-coated HPTLC plates with polymeric binders, are sufficiently hard so as not to be easily damaged by the capillary tubes used for sample application.

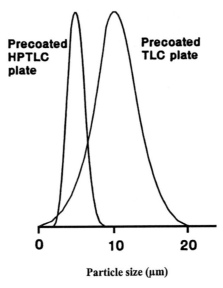

Figure 40 *Comparison of particle size distribution for silica gel used in TLC and HPTLC pre-coated plates*

The smaller particles, similar in size and quality to HPLC packing materials, give a lower theoretical plate height (H) and hence higher efficiency. However, this is only fully utilised if the plates are not overloaded with too much sample, the spot size is kept small (about 1.0 mm), and the plate is developed only to the extent necessary for complete resolution (often only 5 cm and rarely more than 8 cm). A direct comparison of theoretical plates in HPTLC with HPLC serves little purpose as the number found is only valid for the spot used for calculation. The basic problem is that all analytes do not travel the same migration distance and are not measured in retention time as in column chromatography. In planar chromatography, separations are in distance, spots and bands will broaden to differing extents, and hence efficiency can only be constant for a specific chromatogram. Unfortunately the nature of TLC causes a chaotic situation in which the conditions usually vary with time during development. It is therefore difficult to attempt a theoretical model that will fit the separation exactly. However, it is possible to make good approximations. Spots chosen for measurement should be compact and symmetrical, such that a scanner would record a Gaussian peak. This would only be possible for analytes with an R_f value between 0.1 and 0.9, (see Chapter 4 for measurement of R_f). Outside of this range there would be too much shape distortion of the spots either near the origin or solvent front.

Chromatographic spreading (diffusion and migration of components during development) is characterised by the height of an equivalent theoretical plate (HETP), H, which is give by the Knox equation:

$$H = \frac{B}{u} + Au^{1/3} + Cu \qquad (1)$$

where u is the solvent velocity and A, B, and C are experimentally found coefficients.

This equation was first proposed by Knox for column chromatography. A more general approach involves the use of dimensionless reduced values:

$$h = \frac{b}{v} + av^{1/3} + cv \qquad (2)$$

This equation describes a curve with a minimum point.

The "a" term takes into account the local flow velocities due to the size and shape of the sorbent particles.

The "b" term is the molecular diffusion of the solute in the solution, (remember when the mobile phase migration is halted, diffusion will still continue). The "c" term takes account of delays caused by mass transfer processes during sorption and desorption of solute molecules.

In practice, a better quality of sorbent and smaller particle size will result in sharper curves with lower minimums for equation (2) as shown in the example in Figure 41.

HPTLC plates give HETP values of about 12 micrometers and a maximum of about 5000 usable theoretical plates. By contrast, conventional TLC plates have HETP values of about 30 micrometers and approximately 600 theoretical plates.

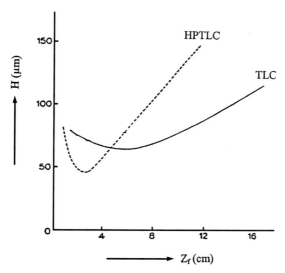

Figure 41 *Plate height (H) versus solvent migration distance (Z_f) comparison between silica gel HPTLC and TLC. Typically demonstrated with a chloroaniline (R_f 0.35) as a standard and using toluene as the mobile phase. As the graph shows, at low Z_f (short development distances) improved resolution is observed for the HPTLC layers, but the effect diminishes with increased development distances*

Table 6 *Comparison of silica gel HPTLC and TLC chromatographic layers*

	HPTLC layer	*TLC layer*
Particle size	5–6 μm	10–12 μm
Pore diameter	60 Å	40 Å, 60 Å, 80 Å, 100 Å
Plate dimensions	10×10 cm, 10×20 cm, 20×20 cm	5×10 cm, 5×20 cm, 10×20 cm, 20×20 cm
Layer thickness	0.20–0.25 mm	0.20–0.25 mm
Number of samples that can be applied per plate	Up to 75	Up to 16
Spot size recommended	~1 mm	2–5 mm
Spot loading	50–200 nl	1–5 μl
Band size recommended	5–10 mm	10–15 mm
Band loading	1–4 μl	5–10 μl
Sensitivity limit	Upper pg (fluorescence)	ng
Normal development time	2–30 minutes	15–120 minutes

It follows therefore that HPTLC offers an increase in performance that is an order of magnitude greater than TLC. Thus it is possible to carry out separations on HPTLC that were not possible before on TLC plates, and for those where it was possible, to shorten the time of separation dramatically. See Table 6 for comparison of plates.

HPTLC is therefore a more rapid, efficient and sensitive technique than conventional TLC. For *in situ* quantitative analysis using spectrodensitometers, it is essential that HPTLC layers are used for the most reliable results. HPTLC represents a considerable advance in the practice of TLC.

9.1 HPTLC Spherical Silica Gel 60

Most HPTLC silica gel 60 layers are manufactured using silica gel of irregular particle shape. Although these represent a major advance over conventional TLC layers, further improvements in solvent capillary flow characteristics, particle size reduction and sorbent purity can be made with spherical silica gels, similar to those presently available in HPLC columns. One of the major problems with particle size reduction of silica gel is the decrease that results in the developing solvent flow velocity. As the process of TLC development usually depends on capillary flow without the application of external pressure, smaller particle sizes than 5 μm result in long development times. This can be offset to some extent by improved binder characteristics, the use of sorbents with spherical particles and narrower particle size distribution, where a much more regular capillary flow of solvent is possible.

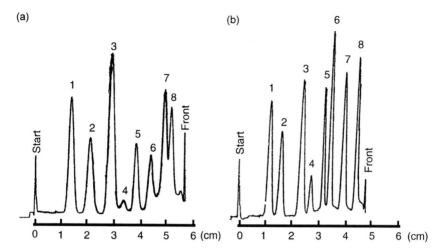

Figure 42 *Comparison between HPTLC silica gel 60, (a) 5–6 μm irregular particles and*
(b) 4 μm spherical particles for the separation of eight pesticides
Mobile phase: petroleum spirit (40–60 °C)/acetone (70 + 30 v/v)
Peaks: 1, hexazinon; 2, metoxuron; 3, monuron; 4, aldicarb; 5, azinphosmethyl;
6, prometryn; 7, pyridate; 8, trifluralin
(By permission of Merck)

Figure 42 demonstrates the improvements that occur in the separation resolution
using a commercially available HPTLC spherical silica gel 60, 4 μm particle size.

The higher purity silica gel in these layers also has the added benefit of lower
background scanning "noise". It is therefore possible to obtain improved detection
limits for spectrodensitometric scanning for quantitative determinations of analytes.
This background "noise" is, in fact, sufficiently reduced such that the TLC
technique can be hyphenated to Raman spectroscopy as a means of further
identification of unknowns or of quantification, (see Chapter 8 where the value of
this hyphenated technique is illustrated).

10 Improving Resolution with Buffers and Complexing Agents

10.1 Impregnation of Sorbent before Layer Coating

In this procedure the silica gel slurry that is normally used to prepare the layer
coating for application on to the glass backing is initially pretreated with the
impregnation reagent. Solutions of buffers, complexing reagents, acids, bases, salts,
or water-soluble organic compounds are effective for this type of impregnation
technique. Once the modified slurry has been prepared it can then be coated on to
the backing plate in the usual way. The technique is limited to water-soluble agents.
Because preparation of TLC plates in the laboratory is now rarely done, this is not a
popular technique.

10.2 Impregnation of the Ready Coated Layer

This can be performed in three ways:

1. Immersion of the pre-coated plate in a solution (usually 5–10%) of the impregnating agent dissolved in a suitable volatile solvent. The solvent is then evaporated off either at ambient or elevated temperature.
2. Spraying a solution of the impregnation reagent on to the plate and then removing the solvent as before.
3. Blank development of the TLC plate in a chromatography tank using the solution of the impregnation reagent as the developing solvent. Migration would be allowed to proceed until the solvent front had reached the top of the layer, and for some time after this so that the "real" solvent front had migrated sufficiently. The solvent is then removed as before.

At times acidic silica gel layers have been used to separate reactive acidic compounds (phenols, acids)[79] and basic layers for the separation of alkaloids, and amines.[80] The acid impregnation was achieved with 0.05–0.25 M oxalic acid and the base used was potassium hydroxide. To ascertain how an unknown compound would react, a pH gradient plate could be used where the silica gel layer is applied to the glass backing surface, starting with low pH at one edge and spreading across the layer to high pH at the opposite edge. The sample would be applied at appropriate intervals along the lower edge. After development in an appropriate solvent the variation of the migration distances could be compared.

Buffers are quite often used in the separation of carbohydrates whether on silica gel 60 or on amino-bonded silica gel layers. Either potassium dihydrogen orthophosphate or the corresponding sodium salt is used (0.2–0.5 M) to inhibit the formation of glycamines from reducing sugars. Mono-, di-, and trisaccharides have been well resolved on silica gel 60 TLC plates previously impregnated with 0.5 M sodium dihydrogen orthophosphate.[81–83] A specific separation can be obtained by impregnation with compounds that will form coordination, chelation or inclusion complexes with the sample components to differing degrees. Boric acid and disodium tetraborate are often employed as impregnation agents for complex formation. These are used for improved resolution in the separation of all types of carbohydrates.[81] EDTA impregnated layers have been used for the separation of anti-microbials.[84] Silver nitrate impregnation is widely used on silica gel 60 and as such merits special consideration.

11 Silver Nitrate Impregnation

Although normal silica gel 60 is a very versatile sorbent for the separation of many aromatic and aliphatic saturated and unsaturated organic compounds, there are occasions when silica gel has insufficient resolving power to achieve adequate separation of unsaturated organic species, particularly where *cis/trans* isomerisation may be involved. Argentation or silver nitrate TLC overcomes this problem in a unique way and enables excellent separations of many types of lipids, fatty acids, olefins, steroids and triterpenes.[85-93]

The separation is based on the known interaction of the silver ion, Ag^+ with ethylenic π-bonds present in the solute molecules. The strength of the interaction will depend on a number of factors:

1. The number of olefinic bonds present.
2. Steric hindrance surrounding the olefinic bonds.
3. The position of the olefinic bonds in the structure.

The stronger the interaction and hence complexation, the more the compound is retained on the layer. The technique is very powerful as it will even allow *cis* and *trans* isomers of some organic compounds to be resolved. The difference in retention observed for such geometric isomers is normally due to the degree of steric hindrance around the double bond. Good examples of this effect are found in the separation of unsaturated fatty acids and lipids. In some cases the silver nitrate can be added to the solvent mixture used for development (where aqueous-based solvents are used). This will be described later in Chapter 5. However, in by far the majority of cases the silver nitrate is impregnated on to the layer prior to development. This is achieved in the following way:

Dip the silica gel pre-coated TLC/HPTLC plate in an aqueous solution of silver nitrate (20% w/v) for 15–20 minutes. Then in the absence of light, dry the plate in air, and finally activate in an oven at 80 °C for about 30–60 minutes. Using this procedure, approximately 1.7 g silver nitrate will be taken up by a 0.25 mm thick layer, giving a concentration of silver nitrate after impregnation of about 40%.

A similar procedure can be used with reversed-phase silica gel plates. The solutions of silver nitrate required are normally of lower strength (1–10% w/v) and are prepared in alcohol/water mixtures to improve "wetting" of the layer. The plate is then dried in the same way as for normal-phase silica gel. Sample application and chromatographic development is performed in exactly the same way as for unimpregnated plates. An example of the application of this technique is shown in Figure 43.

12 Charge Transfer TLC

Charge transfer TLC is a unique technique that relies on an impregnating agent in the chromatographic layer that acts as a π-electron acceptor. These π-acceptors are usually aromatic, unsaturated alicyclic or heterocyclic organic molecules with functional groups of high electron affinity. The following different electron acceptors have been used for impregnation: 2,4,7-trinitrofluorenone, 2,4,6-trinitrophenol (picric acid), 1,3,5-trinitrobenzene, benzoquinone, tetramethyl uric acid, pyromellitic dianhydride, sodium desoxycholate, urea, nucleic acid bases, amino-acids and caffeine.[94-98]

The industrial application that has proved to be of major significance in this area is the separation of polyaromatic hydrocarbons using a caffeine impregnated HPTLC silica gel plate (particularly for the detection and quantification of low

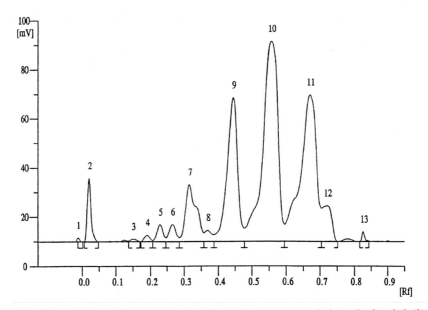

Figure 43 *Separation of unsaturated triacylglycerides on reversed-phase C_{18} bonded silica gel impregnated with silver nitrate. Triglycerides identified all have three C_{18} chains, but vary in the number of double bonds*
Sample: soybean oil
Mobile phase: methanol/dichloromethane/ethyl acetate/acetic acid/water (36 + 26 + 20 + 12 + 6 v/v)
Major Peaks: 7, C_{18} (211); 9, C_{18} (221); 10, C_{18} (222); 11, possibly C_{18} (223)
Detection: visualisation with molybdophosphoric acid reagent. Scanned at 700 nm using CAMAG Scanner 3

levels of polyaromatic hydrocarbons in water supplies). The caffeine acts as a π-electron acceptor and the polyaromatic hydrocarbons exhibit to varying degrees the ability to be π-electron donors. A charge transfer complex is thus formed, (see Figure 44).

The preparation of the impregnated HPTLC plates is a simple operation involving the dipping of the pre-coated HPTLC plate in a solution of caffeine (4 g) in chloroform (96 g) for a few seconds (4 s) and then drying in the oven (110 °C for 30 min.). For quantitative determinations the plates can be pre-washed if required by running a blank chromatogram with dichloromethane as mobile phase. Of course, reactivation as before will then be necessary. (Commercially, caffeine impregnated plates are directly available from Merck and Macherey-Nagel.)

The separation is performed in a standard chromatography chamber. But in order to achieve good resolution of the charge transfer complexes, the separation needs to be carried out at the low temperature of −20 °C. The polyaromatic hydrocarbons are detected by the strong fluorescence they exhibit when exposed to UV radiation. To stabilise and enhance this fluorescence, the developed HPTLC plates can be dipped in a liquid paraffin – n-hexane mixture. Quantification is carried out by scanning the plate for fluorescence intensity at 366 nm, (see Figure 45).

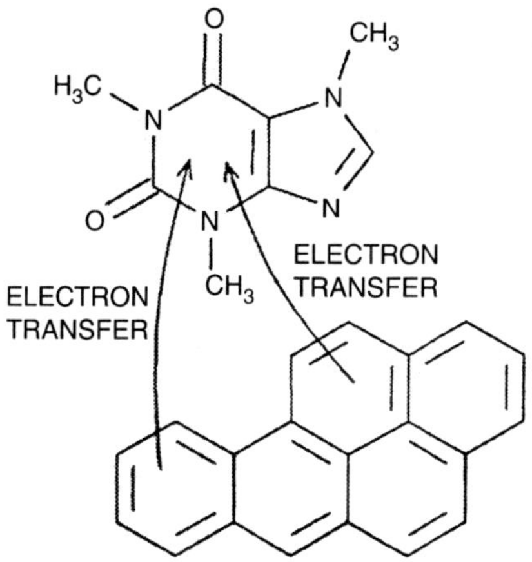

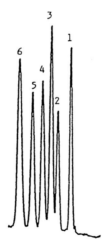

Figure 44 *Formation of a charged transfer complex between benzo[α]pyrene and caffeine. The π-electrons from the aromatic structure of the analyte are "accepted" by the cyclic caffeine molecule*

Figure 45 *Separation of polyaromatic hydrocarbons on caffeine impregnated HPTLC silica gel plates*
Mobile phase: di-isopropyl ether/n-hexane (4 + 1 v/v)
Temperature: −20 °C
Detection: Fluorescence at 366 nm
Peaks: 1, benzo[ghi]perylene; 2, indeno[123-cd]pyrene; 3, benzo[α]pyrene; 4, benzo[b]fluoranthene; 5, benzo[k]fluoranthene; 6, fluoranthene 2 ng/spot (1–5) and 10 ng/spot (6)
(W.Funk, *et al., J. Planar Chromatography,* 1989, **2**, 31, by permission of the publisher)

As there are many molecules which can act as π-donors or π-acceptors the potential for separations on thin layers must be great particularly since the impregnation procedure is so quickly and easily accomplished. Although most impregnation of TLC layers has been with π-electron acceptors, there is obviously also potential for π-electron donor molecules to be used in a similar way.

13 Preparative Layer Chromatography (PLC)

Preparative layers normally refer to TLC layers that are 0.5 mm or more thick. As a general rule the maximum practical thickness is 2 mm although attempts have been made to work with layers 10 mm thick. Manufacturers supply pre-coated PLC plates that are reasonably abrasive resistant. PLC has been reviewed on a number of occasions in the literature indicating that it is still a popular technique.[99–101] There are three main differences between the technique of PLC and conventional TLC:

1. The sample is always applied in a band, preferably across most of the width of the plate to allow as high a loading as possible on the layer.
2. Detection of separated substances is almost exclusively by UV absorbance or fluorescence/fluorescence quenching.
3. Usually multiple development is required to obtain sufficient resolution of separated sample components.

As larger volumes are applied in PLC compared with TLC, the use of automated band application equipment as described later in Chapter 4 is essential for accurate and reliable results. Sample solutions can be applied as a band across the full width of the PLC plate. This enables the maximum amount of sample possible to be applied (volumes of $\sim$500 μl maximum in one dose can be applied using automated equipment). However, it is important to allow about a 2 cm gap between the end of the band and the edge of the plate. This will avoid any "edge effects" that can occur during development due to layer thickness variations of the sorbent at the layer edges. The thickness of the layer and the ability to apply samples across the width of the plate mean that milligram to very low gram loadings can be applied, but unfortunately long development times are unavoidable by the use of normal capillary forces. Often a separation that takes about 30–60 minutes on TLC will take several hours on a 2 mm thick PLC layer. This is not necessarily a disadvantage of the technique as separations can often be run overnight as no active part needs to be taken by the chromatographer during development. Normally the eluent chosen will have been determined from previous TLC experiments. Development on the PLC plate can be carried out several times (usually 3 to 5 times) if required with intermediate drying. Resolution is often improved in this way. Often it proves advantageous to use a solvent mixture as mobile phase that is slightly less polar than that used in the TLC. On the first development the compounds separated should only be allowed to migrate about 2 cm. On the second and subsequent developments the polarity of the mobile phase can be slightly increased to improve resolution.

In PLC it will be necessary to remove the compounds of interest for further analysis or use. A small, sharp bradawl can be used to mark out the position of the zones on the layer. Always remember to mark slightly below the zones as the migration rate of separated compounds will be slightly slower at the glass surface compared with that at the outer surface of the silica gel. The zones can then be carefully removed with a metal spatula and various suction devices. An appropriate solvent is used to solubilise the analytes. The sorbent can then be separated by filtration and the eluent concentrated by evaporation to yield the compounds of interest.

14 References

1. R.M. Scott, *J. Chromatogr.*, 1973, **11**, 129.
2. K.K. Unger, 'Porous Silica its Properties and Use as Support in Column Liquid Chromatography', in *Journal of Chromatography Library*, vol. 16, K.K. Unger (ed), Elsevier, Oxford, UK, 1979, 1–2.
3. R.P.W. Scott in *Silica Gel and Bonded Phases*, R.P.W. Scott and C.F. Simpson (eds), J. Wiley, Chichester, UK, 1993, 23–27.
4. H.R. Felton in *Moisture and Silica Gel* (Technical Report), Analtech, Newark, USA, 1979, 7905.
5. K.R. Lange, *J. Colloid Sci.*, 1965, **20**, 231.
6. S. Gocan in *Modern Thin-Layer Chromatography*, N. Grinberg (ed), Marcel Dekker Inc., New York, USA, 1990, 11.
7. K.Y. Lee, D. Nurok, and A. Zlatkis, *J. Chromatogr.*, 1979, **174**, 187.
8. J.C. Touchstone, R.E. Levitt, R.D. Soloway, and S.S. Levin, *J. Chromatogr.*, 1979, **178**, 566.
9. H. Scherz, G. Stehlik, E. Baucher and K. Kaindl, *Chromatogr. Rev.*, 1968, **10**, 1.
10. M. Ghebregzabher, S. Rufini, B. Monaldi and M. Lato, *J. Chromatogr.*, 1976, **127**, 133.
11. C. Radecka and W.L. Wilson, *J. Chromatogr.*, 1971, **57**, 297.
12. L. Lepri, P.G. Desideri and D. Heimler, *J. Chromatogr.*, 1978, **155**, 119–127.
13. H.K. Mangold and R. Kammereck, *J. Amer. Oil Chemists Soc.*, 1962, **39**, 201.
14. E. Stahl, *Arch. Pharm.*, 1959, **292**, 411.
15. M.S.J. Dallas and F.B. Padley, *Lebensm. Wiss. U-Technol.*, 1977, **10**, 328–331.
16. H.K. Blat and G.A.S. Ansari, *J. Chromatogr.*, 1989, **483**, 369–378.
17. G.B. Proctor and R.W. Horobin, *Stain Tech.*, 1985, **60**, 1–6.
18. R.E. Kaiser and R. Rieder, *J. Chromatogr.*, 1977, **142**, 411–420.
19. D.W. Armstrong, K.H. Bui and R.E. Boehm, *J. Liq. Chromatogr.*, 1983, **6(1)**, 1–22.
20. D.W. Armstrong and K.H. Bui, *J. Liq. Chromatogr.*, 1984, **7(1)**, 45–58.

21. J. Sherma, D.W. Armstrong and B.P. Sleckman, *J. Liq. Chromatogr.*, 1983, **6(1)**, 95–108.
22. M. Vanhaelen and R. Vanhaelen-Fastré, *J. Chromatogr.*, 1980, **187**, 255–260.
23. B. Amidzhin and B. Nikolova-Dansyanova, *J. Chromatogr.*, 1988, **446**, 259–266.
24. J. McSavage and P.E. Wall, *J. Planar Chromatogr.*, 1998, **11**, 214–221.
25. D. Giron and P. Groell, *J. HRC and CC*, 1978, 67–68.
26. G.P. Ellis and J. Honeyman in *Advances in Carbohydrate Chemistry*, vol. 10, M.L. Wolfrom (ed), Academic Press, New York, USA, 1955, 95.
27. W. Jost and H.E. Hauck in *Instrumental High Performance Thin-Layer Chromatography* (Interlaken 1982), R.E. Kaiser (ed), Institute for Chromatography, Bad Dürkheim, Germany, 1982, 25–37.
28. W. Jost and H.E. Hauck, *J. Chromatogr.*, 1983, **261**, 235–244.
29. W. Jost and H.E. Hauck, *Anal. Biochem.*, 1983, **135**, 120–127.
30. R. Klaus, W. Fischer and H.E. Hauck, *Chromatographia*, 1990, **29**, 467–472.
31. M. Ericsson and L.C. Blomberg, *J. HRC and CC*, 1980, **3**, 345.
32. W. Jost and H.E. Hauck in *Instrumental High Performance Thin-Layer Chromatography* (Wurzberg 1985), R.E. Kaiser (ed), Institute for Chromatography, Bad Dürkheim, Germany, 1985, 83–91.
33. J.S. Kang and S. Ebel, *J. Planar Chromatogr.*, 1989, **2**, 434–437.
34. H.E. Hauck, M. Mack, S. Reuke and H. Herbert, *J. Planar Chromatogr.*, 1989, **2**, 268–275.
35. W. Jost and H.E. Hauck in *Instrumental High Performance Thin-Layer Chromatography* (Selvino/Bergamo 1987), R.E. Kaiser, H. Traitler and A. Studer (eds), Institute for Chromatography, Bad Dürkheim, Germany, 1987, 241–253.
36. L. Witherow, R.J. Thorp, I.D. Wilson and A. Warrander, *J. Planar Chromatogr.*, 1990, **3**, 169–172.
37. I.W. Wainer, C.A. Brunner and T.D. Doyle, *J. Chromatogr.*, 1983, **264**, 154.
38. W.H. Pirkle, D.W. House and J.M. Finn, *J. Chromatogr.*, 1980, **192**, 143–158.
39. W.H. Pirkle, J.M. Finn, J.L. Schreiner and B.C. Hamper, *J. Am. Chem. Soc.*, 1981, **103**, 3964–3966.
40. P.E. Wall, *J. Planar Chromatogr.*, 1989, **2**, 228–232.
41. P.E. Wall in *Instrumental High Performance Thin-Layer Chromatography* (Brighton, 1989), R.E. Kaiser (ed), Institute for Chromatography, Bad Dürkheim, Germany, 1989, 237–243.
42. N. Oi, H. Kitahara, T. Doi and S. Yamamoto, *Bunseki Kagaku*, 1983, **32**, 345; 1983, **99**, 817906.
43. C.A. Brunner and I. Wainer, *J. Chromatogr.*, 1989, **472**, 277–283.
44. S.M. Hau and D.W. Armstrong in *Planar Chromatography in the Life Sciences*, J.C. Touchstone (ed), J. Wiley, Chichester, UK, 1990, 87–89.
45. A. Alak and D.W. Armstrong, *Anal. Chem.*, 1986, **58**, 582.
46. V.A. Davankov, A.S. Bochkov and A.A. Kurganov, *Chromatographia*, 1980, **13**, 677.
47. V.A. Davankov, A.S. Bochkov and Y.P. Belov, *J. Chromatogr.*, 1981, **218**, 547–557.

48. K. Günther, J. Martens and M. Schickedanz, *Angew. Chem.*, 1984, **96**, 514.
49. M. Mack, H.E. Hauck and H. Herbert, *J. Planar Chromatogr.*, 1988, **1**, 304–308.
50. M. Mack and H.E. Hauck, *Chromatographia*, 1988, **26**, 3–11.
51. U.A. Th. Brinkman and D. Kamminga, *J. Chomatogr.*, 1985, **330**, 375–378.
52. J.G. Heathcote and C. Haworth, *J. Chromatogr.*, 1969, **43**, 84–92.
53. J.G. Heathcote and C. Haworth, *J. Chromatogr.*, 1969, **41**, 380–385.
54. R.W. McBride, D.W. Jolly, B.M. Kadis and T.E. Nelson, *J. Chromatogr.*, 1979, **168**, 290–291.
55. K. Jones and J.G. Heathcote, *J. Chromatogr.*, 1966, **24**, 106.
56. J.G. Heathcote, D.M. Davies, C. Haworth and R.W. Oliver, *J. Chromatogr.*, 1971, **55**, 377–384.
57. J.M. Davies, *J. Chromatogr.*, 1972, **69**, 333–339.
58. K. Randerath and H. Struck, *J. Chromatogr.*, 1961, **6**, 365.
59. A. Mohammad, S. Tiwari and J.P. Chahar, *J. Chromatogr. Sci.*, 1995, **33**, 143–147.
60. K. Randerath in *Thin-Layer Chromatography*, Academic Press, London, UK, 1963, 195–197.
61. L. Lepri, V. Coas, P.G. Desideri and A. Zocchi, *J. Planar Chromatogr.*, 1994, **7**, 376–381.
62. H. Brockmann and H. Schodder, *Ber. Dtsch. Chem. Ges.*, 1941, **74**, 73.
63. H. Rössler in *Thin-Layer Chromatography A Laboratory Handbook*, E. Stahl (ed), Springer-Verlag, Berlin, Germany, 1969, 28.
64. R. Bushan, *J. Chomatogr. Sci.*, 1991, **55**, 353–387.
65. E. Soczewiński and H. Szumilo, *J. Chromatogr.*, 1973, **81**, 99–107.
66. I.S. Bhatia, J. Singh and K.L. Bajaj, *J. Chromatogr.*, 1973, **79**, 350–352.
67. R.S. Bayliss, J.R. Knowles and G.B. Wybrandt, *J. Biochem.*, 1969, **113**, 377–386.
68. H. Rössler in *Thin-Layer Chromatography A Laboratory Handbook*, E. Stahl (ed), Springer-Verlag, Berlin, Germany, 1969, 29.
69. J.K. Rozylo, D.G. Chomicz and I. Malinowska in *Instrumental High Performance Thin-Layer Chromatography* (Würzburg 1985) R.E. Kaiser (ed), Bad Dürkheim, Germany, 1985, 173–187.
70. H. Rössler in *Thin-Layer Chromatography A Laboratory Handbook* E. Stahl (ed), Springer-Verlag, Berlin, Germany, 1969, 40–41.
71. H. Egan, E.W. Hammond and J. Thomson, *Analyst*, 1964, **89**, 480.
72. O. Crépy, O. Judas and B. Lachese, *J. Chromatogr.*, 1964, **16**, 340–344.
73. Z. Gregorowicz, J. Kulicka and T. Suwinka, *Chem. Anal.*, 1971, **16**, 169.
74. J.A.W. Gosselé, *J. Chromatogr.*, 1971, **63**, 433–437.
75. H. König and M. Schüller, *Z. Anal. Chem.*, 1979, **294**, 36.
76. J. Sherma in *Practice and Applications of Thin Layer Chromatography on Whatman KC_{18} Reversed Phase Plates*, TLC Technical Series, vol. 1, Whatman Chemical Separations Inc., New Jersey, USA, 1981, 10.
77. D.C. Abbott and J. Thomson, *Chem. Ind.*, 1965, 310.

78. A. Musgrave, *J. Chromatogr.*, 1969, **41**, 470.
79. M.S.J. Dallas, *Nature*, 1965, **207**, 1388.
80. E. Stahl, *Arch. Pharm.*, 1959, **292**, 411.
81. M. Ghebregzabher, S. Rufini, B. Monalgi and M. Lato, *J. Chromatogr.*, 1976, **127**, 133–162.
82. K.Y. Lee, D. Nurok and A. Zlatkis, *J. Chromatogr.*, 1979, **174**, 187–193.
83. R. Klaus and J. Ripphahn, *J. Chromatogr.*, 1982, **244**, 99–124.
84. C. Radecka and W.L. Wilson, *J. Chromatogr.*, 1971, **57**, 297.
85. E.W. Hammond in *Chromatography for the Analysis of Lipids*, CRC Press, Florida, USA, 1993, 39–54.
86. F.D. Gunstone in *Fatty Acid and Lipid Chemistry*, Blackie Academic & Professional, London, UK, 1996, 18–19.
87. J.C. Touchstone, *J. Chromatogr. B*, 1995, **671**, 169–195.
88. G. Dobson, W.W. Christie and B. Nikolova-Damyanova, *J Chromatogr. B*, 1995, **671**, 197–222.
89. R.O. Adlof, *J. Chromatogr. A*, 1996, **741**, 135–138.
90. B. Nikolova-Damyanova, W.W. Christie and B. Herslof, *J. Planar Chromatogr.*, 1994, **7**, 382–385.
91. M. Inomata, F. Takaku, Y. Nagal and M. Saito, *Anal. Biochem.*, 1982, **125**, 197–202.
92. H.K. Bhat and G.A.S. Ansari, *J. Chromatogr.*, 1989, **483**, 369–378.
93. K. Aitzetmüller and L.A.G. Goncalves, *J. Chromatogr.*, 1990, **519**, 349–358.
94. W. Funk, V. Glück, B. Schuch and G. Donnevert, *J. Planar Chromatogr.*, 1989, **2**, 28–32.
95. W. Funk, G. Donnevert, B. Schuch, V. Glück and J Becker, *J. Chromatogr.*, 1989, **2**, 317–320.
96. J. Tríska, N. Vrchotová, I. Šafařik and M. Šsfsříková, *J. Chromatogr. A*, 1998, **793**, 403–408.
97. G.D. Short and R. Young, *Analyst*, 1969, **94**, 259.
98. V. Libíčková, M. Stuchlík and L. Krasnec, *J. Chromatogr.*, 1969, **45**, 278.
99. H. Halpaap in *Chromatographic and Electrophoretic Techniques*, vol. 1, I. Smith (ed), Interscience, New York, USA, 1969, 834–886.
100. H.R. Felton in *Advances in TLC – Clinical and Environmental Applications*, J. Touchstone (ed), Wiley-Interscience, New York, USA, 1982, 13–20.
101. Sz. Nyiredy in *Handbook of Thin Layer Chromatography*, 2nd edn, J. Sherma and B. Fried (eds), Marcel Dekker, New York, USA, 1996, 307–340.

CHAPTER 3

Sample Pre-treatment

1 Introduction

Pre-treatment of samples before application to the TLC/HPTLC layer rarely requires more than a few basic steps. The quality of commercially available TLC and HPTLC layers is such that most analytes can be more easily distinguished from impurities than was possible with "home made" plates. Usually the presence of sample contaminants does not cause a problem in TLC. Whereas the repeated injection of contaminants on to an HPLC column can quickly render it useless, TLC or HPTLC plates are normally only used once and are generally less sensitive to contamination. When the impure sample is applied to the sorbent layer as a spot or band, both the components of interest and the interfering impurities are deposited together. Once development begins, the contaminants are often left behind at or near the origin whilst the components of interest migrate in the direction of flow of the solvent front. If the sample solvent is mainly aqueous or viscous, then dilution with an organic polar solvent, like methanol, ethanol or acetonitrile, will aid application to the layer. The sample solution is then able to wet the surface and penetrate into the sorbent effectively. The application will take on a regular spherical shape in the case of spots and a fine, sharp, well-defined line in the case of bands. Filtering of the sample solution is also an important step that can improve the result of the eventual chromatogram. As the sample volumes will be small, simple syringe filters of 0.45 or 0.2 μm pore size will suffice. A variety of such filters are commercially available, based on cellulose, cellulose acetate and nitrate, alumina PES, polypropylene or PTFE.

In cases where the contaminants do interfere with the TLC chromatogram of the components of interest, then sample clean-up procedures of the type commonly used in HPLC can be used. These involve the use of solid phase extraction (SPE) cartridges and an appropriate vacuum manifold. Solid phase extraction sorbents can be highly specific, exploiting the different physico-chemical properties of the sorbents whilst others are of a fairly simple nature. These currently include diatomaceous earth, silica gel, C_2, C_8, C_{18}, CN, Diol, NH_2 and phenyl-bonded silica gel, and cation-and anion-exchangers based on silica and various polymers. Often very selective extractions can be achieved by choosing a sorbent that either absorbs the analyte but not the impurities, or absorbs the impurities and allows elution of the analyte. Where the analyte is absorbed, the extraction column can be washed with a solvent that removes any remaining impurities, but has no effect on the analyte. By careful choice of solvent the analyte is finally eluted, and concentrated if required prior to application to the TLC plate. It is also possible to use these SPE columns for sample concentration and trace enrichments. Biological samples, for example plasma, can be treated prior to chromatography with trichloroacetic acid, perchloric acid or acetonitrile in order to remove protein material by precipitation. However, a separate sample pre-treatment step can often be eliminated by pre-derivatisation of the analytes in a sample, and the use of concentration zone TLC plates. This is described more fully in Chapters 2 and 6, respectively.

2 Extractions from Solid Samples

Samples from plant material including herbal products sometimes require more than simply shaking the product with an appropriate solvent to release the analytes of interest. For example, leaves, flowers, and roots often contain a number of classes of compound that differ in polarity. Extraction procedures are therefore necessary to isolate all of the compounds present into their respective groups. An example of a general procedure that can be used in most instances is shown in Figure 1. For the analysis of herbal products for non-polar classes of compounds, such as essential oils, dichloromethane, diethyl ether or n-hexane

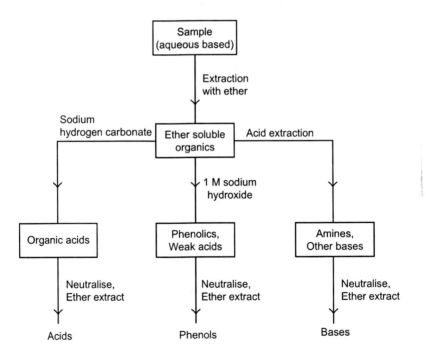

Figure 1   *A general extraction procedure for isolating major groups of compounds from an aqueous based sample*

have been used. Sometimes cold extraction is sufficient, but more usually boiling or Soxhlet extraction in the solvent proves most effective. Most other herbal products containing active polar constituents can be boiled in methanol, or sometimes ethanol, or dilutions of either solvent in water, to completely solubilise the analytes of interest. Further fractionation into different groups of compounds based on polarity can be carried out if so desired.[1] The use of pH adjustments of ether extracts can prove most effective. At high pH using dilute sodium hydroxide solution or sodium bicarbonate, organic acids and phenolics can be converted into salts and hence can then be solubilised in aqueous conditions. At low pH using dilute mineral acids, amines and amides will form salts and can then also be extracted in similar fashion. The aqueous extracts are neutralised and the organic species extracted with a suitable immiscible solvent (*e.g.* diethyl ether), and concentrated by evaporation ready for application to the TLC layer.

Samples of microbiological origin sometimes require special treatment. Organisms can be separated from the nutrient medium by filtration or centrifugation. Homogenisers or sonicators are often used to break down the cellular material for further analysis. Animal or human body fluid samples are best pre-treated using solid phase extraction systems described in the next section.

3 Solid Phase Extraction Systems (SPE)

3.1 Diatomaceous Earth

Diatomaceous earth is a natural siliceous deposit of diatom frustules. Kieselguhr, celite, bentonite and fuller's earth are all varieties of diatomaceous earth, but these are normally used as filtering aids or clearing agents. However, certain highly purified grades of kieselguhr are used as absorption materials for sample pre-treatment in chromatography, and indeed as we have seen in the previous chapter as TLC sorbents. Short columns containing a few grams of dry sorbent can easily be prepared using a glass or plastic tube with a taper at one end. A plug of glass wool can act as an effective filter if a commercially manufactured tube is not available. Usually the sample is poured directly into the column, and left for a period of 10–15 minutes. A non-water miscible solvent is then used to elute the analytes. The eluate can be evaporated to low volume or to dryness and reconstituted in a suitable solvent for application to the TLC layer. Table 1 shows one example using a commercial trade-named product for a purified diatomaceous earth. This type of SPE is often used for the pre-treatment of human or animal body fluids.[2,3]

3.2 Bonded Silica Gels and Polymers

These highly specific bonded phases have been developed particularly for sample pre-treatment in HPLC and GC. They are designed to cope with a wide variety of samples from a wide range of origins.[4–9] However, if needed their properties can be utilised for sample preparation prior to planar chromatography. A number of

Table 1 *Extraction and subsequent TLC procedure for caffeine in tea using a diatomaceous earth (Extrelut^a) as the solid phase extraction material*

Extraction procedure	
Sample:	Tea leaves
Extraction sorbent:	Extrelut$^®$ 3 pre-packed column (Merk, Darmstadt, Germany)
Extraction solvent:	Dichloromethane
Extraction method:	Tea (50 mg) is boiled in water (7 ml) for 2 minutes. After cooling, the liquid is filtered and the volume made up to 10 ml. Polyamide powder (200 mg) is mixed in and after a few minutes, the whole is re-filtered. The extract (3 ml) is applied to the Extrelut column and allowed to absorb for 10 minutes. 3 aliquots of dichloromethane (3 ml of each) are then used to elute the compounds of interest. The eluates are combined.
TLC conditions	
Sorbent:	TLC pre-coated aluminium oxide F_{254} glass plate
Developing solvent:	methanol/water (92:8% v/v)
Sample:	1 μl loadings of the dichloromethane eluate
Detection:	UV at 254 nm
R_f values:	caffeine (0.70), nicotinamide (0.60), trigonelline (0.34), and nicotinic acid (0.08)

^{a} Trade name of Merck KGaA, Darmstadt, Germany

commercial products are available as pre-packed columns containing typically 100–500 mg of sorbent. Some of these are specifically designed for environmental analysis where a high capacity for trace enrichment from water is required.[10]

In normal use a series of steps are taken to attain the best results. Choosing a suitable SPE cartridge for a particular clean-up requires some forethought. The type of cartridge is selected on the basis of the retention it will have for the analytes in the sample, the sample matrix, and the composition of the sample. Non-polar compounds in aqueous solutions or buffers are extracted using reversed-phase sorbent cartridges. The eluents selected are polar in character, such as methanol, acetonitrile, or either solvent mixed with a suitable buffer solution. When the opposite scenario is the case and the sample analytes are polar in character and the sample solvent is non-polar (n-hexane, iso-octane or dichloromethane), a normal-phase SPE cartridge is used. Diol-bonded silica gel is usually best, but cyano-, amino- or silica gel 60 can also be used. Eluents selected are normally of a medium polarity such as acetone, ethyl acetate or tetrahydrofuran, although sometimes lower polarity solvents are sufficient. If the sample contains analytes with positively charged functional groups and the sample solution is aqueous, cation-exchange SPE is the cartridge of choice. Cation-exchange SPE cartridges are based on a sulfonic or carboxylic acid as the functional group. Elution is effected using buffers based on citrate, acetate or phosphate. Conversely, if the sample contains analytes with negatively charged functional groups, such as organic acids in aqueous solution, then an anion-exchange SPE cartridge is used based on an amino or quaternary amine functional group. As before phosphate or acetate buffer solutions are used as eluents. It is important in both the latter cases that the ionic strength of the sample solution is low to obtain maximum ion-exchange capacity in the cartridge. Strong solutions will result in ion leakage.

3.3 Sorbent Conditioning

The support material is first conditioned with an organic solvent (usually acetonitrile or methanol). A single column volume normally suffices. This solvates the hydrocarbon chains bonded to the support. Following this a column volume of the solvent used to dissolve the sample is passed through the SPE cartridge.

3.4 Sample Injection

The sample solution is then forced through the cartridge usually employing a low vacuum. The analytes concentrate on the sorbent, whilst it is hoped that any undesirable components of the solute matrix are not adsorbed and elute from the cartridge. The volume of sample solution used varies, depending on the concentration of the components of interest and on the sensitivity of detection. Although small sample volumes of less than a millilitre are often applied to the SPE cartridge, volumes of 100 ml or more can be used where it is necessary to concentrate the analytes on the top of the sorbent.

3.5 Cleaning

For reversed-phase cartridges, water or sometimes buffer or even water/methanol mixtures are used to wash out any further undesirable components including any left behind in the previous step. Ion-exchange SPE cartridges can be treated in a similar way. Again buffers and buffer/methanol mixtures can be used (up to 10% v/v methanol), but it is important that the pH is maintained at the same value as the original sample solution. Changes in pH may cause premature elution of analytes.

3.6 Recovery of Analytes

A suitable solvent is selected to elute the analytes from the sorbent. What decides the choice of solvent is how well it will weaken the strength of the interaction between the analytes and the sorbent, such that the analytes are eluted, but also leave behind the more strongly bound matrix components. The eluate containing the analytes can be concentrated prior to application onto the TLC layer. Sometimes it is advantageous to evaporate this eluate to dryness and then to reconstitute the sample in a solvent that gives good "wetting" of the TLC layer. However, the choice of solvent here is a very important consideration in planar chromatography and it is recommended that the reader refers to Chapter 4 on this subject. Whatever approach is applied to SPE, it is important that the eluate volume is small enough such that sufficient analyte is present to allow adequate visualisation sensitivity after development of the chromatogram. Conversely it can also be possible that the concentration is so high that application of the sample solution to the TLC plate results in overloading problems during development. In this instance simple dilution of the sample solution is all that is required.

4 References

1. J.B. Harborne in *Phytochemical Methods – A Guide to Modern Techniques of Planar Analysis,* 2nd edn, Chapman and Hall, London, UK, 1984.
2. J. Breiter, *Kontakte* (E Merck), 1981, **2**, 21–32.
3. J. Breiter, R. Helger and H. Lang, *Forensic Science*, 1976, **7**, 131–140.
4. J.P. Lautie and V. Stankovic, *J. Planar Chromatogr.*, 1996, **9**, 113–115.
5. K. Jacob, E. Egeler, B. Hennel and D. Neumeier, *Fresenius-Z. Anal. Chem.*, 1988, **330(4–5)**, 386–387.
6. Y. Ikai, H. Oka, N. Kawamura, M. Yamada, K. Harada and M. Suzuki, *J. Chromatogr.*, 1987, **411**, 313–323.
7. S. Kessel and H.E. Hauck, *Chromatographia*, 1996, **43(7–8)**, 401–404.
8. T. Imrag and A. Junker-Buchheit, *J. Planar Chromatogr.*, 1996, **9(2)**, 146–148.
9. W. Fischer, O. Bund and H.E. Hauck, *Fresenius-J. Anal. Chem.*, 1996, **354(7–8)**, 889–891.
10. A. Junker-Buchheit and M. Witzenbacher, *J. Chromatogr. A.*, 1996, **737**, 67–74.

CHAPTER 4

Sample Application

1 Introduction

Adequate sample preparation and careful application of the sample to the TLC or HPTLC layer is imperative for good chromatographic separations. Frequently poor separations are blamed on the quality of the layer or the technique, rather than the way the sample was processed or applied to the sorbent surface. Without proper attention to detail incorrect sample solvents can easily be chosen resulting in large and/or irregular spot or band shapes for the sample application. Spot or band loadings on the chromatographic layer may also be too concentrated or of different sizes and strengths. Application devices such as pipettes, capillaries, *etc.* may damage the sorbent layer as spots or bands are applied. Band application may be uneven and vary in concentration along the band. It is also important that attention be given to the laboratory environment. A constant temperature and humidity should be maintained and the atmosphere kept relatively free of chemical fumes and solvent vapours. These may affect the sample or be adsorbed into the layer and alter the stationary – mobile phase equilibrium later on during development. Any dust or dirt particles that may find their way onto the sorbent layer should be removed with a soft brush or dry air blower. Breathing on the layer should be avoided as moisture from this source can be adsorbed resulting in a change in the activity of the sorbent.

2 Selecting the Sample Solvent

The solvent used to apply the sample to the TLC plate can have a decisive influence on the spot size. Usually, the general nature of the sample dictates to a large extent the solvent to be used. As a general rule the least polar single solvent or mixture of solvents in which the analytes are completely soluble or completely extracted from the sample matrix should be used. This will assist in attaining small spot size even with one complete loading of the sample from 1–5 μl or 50–250 nl on the TLC or HPTLC layer, respectively. A general recommendation would be to avoid solvents that are too polar, for example water, buffers and diols, as they may not completely "wet" the sorbent layer. A characteristic sign of incomplete "wetting" is a star-shaped spot. Solvents with high boiling points, such as butan-1-ol, toluene and dimethylformamide should also be avoided as they are difficult to evaporate from

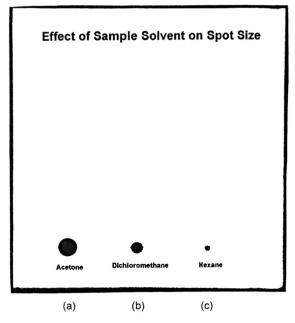

Figure 1 *A mixture of lipophilic dyes dissolved in* (a) *acetone,* (b) *dichloromethane, and* (c) *hexane, deposited at* 1 *μl loadings on a silica gel 60 TLC plate*

the layer surface. Some investigation is required to decide on the most appropriate solvent, one with high volatility and usually low polarity. Optimum spot size will then be possible as demonstrated in Figure 1. However, this cannot be applied as a hard and fast rule as there are cases where a quite polar but highly volatile solvent has proved just as effective as a less polar one.

With the least polar solvent the sample on contact is immediately adsorbed onto the porous surface leaving the solvent to diffuse out into the layer. Conversely, in the case of the more polar solvent, the sample is carried with it and diffuses out into a much larger spot.

Several theoretical attempts have been made to mathematically calculate the expected spot diameter based on the volume of sample applied, the sorbent layer thickness, and the porosity of the sorbent packing. Two of the most useful, the Guiochon-Siouffi and Kaiser equations are now described with the purpose of determining, in a practical way, the volumes of sample that will need to be applied to the sorbent layer to obtain optimum spot size.

3 Theoretical Determination of Spot Size

3.1 Spot Size and Loading Equations

When the sample solution comes into contact with the layer surface, the liquid penetrates into the pores and diffuses into the surrounding sorbent. Assuming no evaporation during this process, the diameter of the sample spot is given by[1]:

$$d = 2 \left[\frac{V}{\pi \times l \times \varepsilon_i (1 - \varepsilon_e)} \right]^{\frac{1}{2}} \tag{1}$$

where d is the diameter of the spot
V is the sample volume
l is the layer thickness
ε_i is the particle porosity of the packing
ε_e is the external porosity of the packing

Practical Calculation

For a dense packing, ε_e is approximately 0.4 and ε_i is typically about 0.65 for most normal- and reversed-phase silica gels. Typically for a 100 nl sample applied as a spot to an HPTLC layer, 0.25 mm thick,

$$d = 2 \left[\frac{100 \times 10^3}{\pi \times 0.25 \times 0.65(1 - 0.4)} \right]^{\frac{1}{2}} = 1.1 \, \text{mm}$$

The value of d could be reduced further by applying a smaller loading, combining this with a finer capillary, perhaps using a less polar solvent, or depositing the sample solution under a steady stream of nitrogen or air to accelerate evaporation. It has been shown that d reduces from 1.4 to 0.45 mm when changing the sample solvent from acetone to n-heptane, without altering the sample volume or capillary diameter.[2]

Kaiser[3] derived a similar equation to Guiochon and Siouffi allowing the calculation of maximum spot size and sample volume in what he describes as "real" situations. The relationship is dependent on a choice of sample solvent, where the R_f (retention factor) = 1. In other words the value is calculated based on the situation where there is no retention of the sample:

$$d_{\text{max}} = \left[\frac{4V(1 + As/Am)}{\pi l} \right]^{\frac{1}{2}} \tag{2}$$

where d_{max} is the maximum diameter of the spot
V is the sample volume applied
As/Am is a ratio of stationary phase to mobile phase in a volume of sorbent filled with mobile phase. For HPTLC this has been determined to be 0.395 (according to Kaiser).[3]
l is the layer thickness.

Practical Calculation

To demonstrate how this can be used
  if $d_{max} = 1$ mm and $l = 0.25$ mm, then,

$$V = \frac{l^2 \times \pi \times 0.25 \times 10^3}{4(1 + 0.395)} = \frac{250 \times \pi}{5.580} = 140\,\text{nl}$$

or if $V = 100$ nl and $l = 0.20$ mm, then,

$$d_{max} = \frac{4 \times 100(1 + 0.395)}{0.2 \times 10^3 \times \pi} = \frac{558}{200 \times \pi} = 0.9\,\text{mm}$$

Once the sample spots have been applied to the sorbent layer and the layer dried, the plate is placed in the developing chamber and solvent migration allowed to proceed. Initially when the solvent comes into contact with the dry sample, a number of interactions occur causing changes in spot shape and size. After a few further millimetres of migration of the solvent, equilibration occurs and spots take on a regular circular shape with a slow diffusion rate. If the initial spot size is well below 1 mm, then during the first few millimetres of migration of the mobile phase the spot size expands rapidly to just over 1 mm. The diffusion rate then slows down to the same as that for initial spots of about 1 mm. Initial sample spots of 1.5 mm diameter have a tendency to converge at the beginning of migration, but then continue at the same diffusion rate as smaller spots. However, they assume an elliptical spot shape on migration. The effect of starting with spot sizes ranging from 0.5 to 1.5 mm is shown in Figure 2. After a short migration distance all sample spots diffuse to a size of about 2.5 mm. To avoid the spot distortion produced by initial spot sizes greater than about 1.1 mm, it is important to apply sample volumes, such that the spot diameter is 1 mm or less. However, with current HPTLC plates there seems little advantage in applying spots much smaller than 1 mm. In practice, spots sizes of about 1 mm normally give the optimum resolution of sample components.

3.2 Spot Capacity Equation

The initial spot size is important as it can have a dramatic effect on the separation number or spot capacity, *i.e.* the number of components resolved between the sample origin and the true solvent front. The relationship between size and separation number is given by the following equation:

$$SN = \left[0.425\sqrt{N_{\text{real}}} \times \left(\frac{b_1 - b_0}{b_1 + b_0} \right) \right] - 1 \tag{3}$$

where SN is the separation number
  N_{real} is the number of theoretical plates for a substance of $R_f = 1$
  b_1 is the extrapolated width of the spot at half height for $R_f = 1$
  b_0 is the extrapolated width of the spot at half height for $R_f = 0$

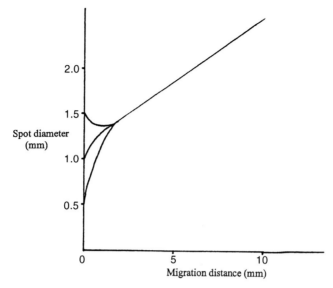

Figure 2 *Typical convergence curves for spot diameters measured in the direction of eluent flow. After the first few millimetres of migration the rate of diffusion settles to a regular pattern*

As b_0, the extrapolated width of the starting spot at half height, approaches zero, the expression $\left(\frac{b_1-b_0}{b_1+b_0}\right)$ approaches unity, giving the maximum value for the separation number. Logically, this means that the smaller the spot size the higher the spot capacity. However, as mentioned above there is little to be presently gained by producing spots with a diameter much below 1 mm, (constraints of the physical characteristics of the layer). Obviously, the potential for higher spot capacity is very dependent on the separation length of the TLC layer. Long plates will apparently allow for a higher separation number. However, the longer the development time and migration distance, the more diffusion of the chromatographic zone that occurs. Hence there is a limit both in time, resolution and sensitivity of detection that can be achieved in increasing the separation number in a single ascending development. For a silica gel 60 TLC layer, spot capacities of 10–20 can be achieved. This can be improved by use of HPTLC layers giving typical values of about 20–25. Further focusing of chromatographic zones is attainable using automated multiple development (see Chapter 5), resulting in separation numbers of up to 30. However, two-dimensional separations, which are unique to planar chromatography, give spot capacities in excess of 250.

The separation number (*SN*) can be determined experimentally in the following way:

1. A mixture of substances, usually dyes (as these are easily detected) is selected such that with the appropriate choice of developing solvent (mobile phase) the components will be separated over the widest range of retention factor (R_f) possible

$$R_f = \frac{\text{Migration distance of substance}}{\text{Migration distance of solvent front from origin}} \qquad (4)$$

The distances measured are shown in Figure 3.

2. After running the chromatogram, the plate is scanned with a spectro-densitometer on an expanded scale to record the peaks obtained.
3. Measurement is made of the peak widths at half height. These values are plotted on an expanded scale against the migration distance (see Figure 4).
4. The points obtained are joined by a straight line and extrapolated to spotwidth b_0 at $R_f = 0$ and b_1 at $R_f = 1$.
5. SN can be calculated by simplifying equation (3) with the expression:

$$N_{\text{real}} = \left(\frac{Z_f}{b_1 - b_0}\right)^2 \times 5.54 \qquad (5)$$

giving:

$$SN = \frac{Z_f}{b_0 + b_1} - 1 \qquad (6)$$

where Z_f is the migration distance of the mobile phase.

As can be noted using these equations, it is possible to calculate the number of theoretical plates (N) and therefore the plate height (H).

From equation (5) we can propose:

$$N = 5.54 \left[\frac{Z_x}{b_x - b_0}\right]^2 \qquad (7)$$

where Z_x is the migration distance of a substance spot, and b_x is the peak width at half height for this spot.

As:

$$H = \frac{Z_x}{N} \qquad (8)$$

$$N = \frac{Z_x}{H}$$

Hence:

$$H = \frac{(b_x - b_0)^2}{5.54 Z_x} \qquad (9)$$

All this shows the importance of a small initial spot size (b_0) that will result in a relatively small final spot (b_x) after development.

This approach recognises the importance of the starting zone dimension on the plate height value (H) and the number of theoretical plates (N). In practice the extrapolated value for b_0 will exceed the value determined spectrodensitometrically for the initial spot before development. The reason for this is the effect explained

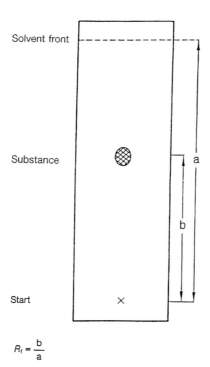

$$R_f = \frac{b}{a}$$

Figure 3 *Measurement of R_f value. Distances are measured for separated components of the sample after development*

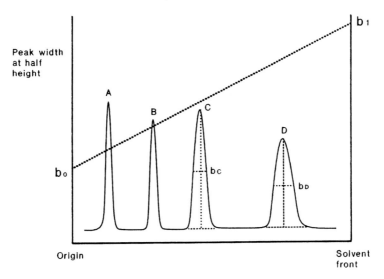

Figure 4 *Typical extrapolation used to determine the b_0 and b_1 values for the calculation of the separation number (SN)*

earlier when the mobile phase first comes into contact with the starting zone (or origin) where the sample has been applied. An initial rapid expansion and reshaping of the sample spot or band occurs. In this region the solvent velocity is high and uneven, and a finite time is required for the sample spot to equilibrate with the mobile phase. The extrapolated value is therefore a more realistic value of the spot dimensions. As was shown earlier, the eventual spot width is independent of the initial spot size where spots are about 1 mm in diameter. The quality of the sorbent layer is the primary consideration for the value of b_0.

4 Sample Loading

Not only is careful control of spot or band size important, but the amount of sample that is loaded on to the TLC/HPTLC layer should be given careful consideration. Layers can easily be overloaded without the result being initially realised. As the overloaded spot or band migrates during development, diffusion into the surrounding sorbent occurs at an abnormal rate particularly in the direction of migration. This results in a "comet-like" streak with tails that vary in length depending on the amount of overloading. Resolution is therefore dramatically reduced. Unfortunately, rather than recognising the source of the problem, the sorbent layer or the mobile phase is often blamed for the apparent poor chromatography. (Usually the streak overlaps other spots in the chromatogram resulting in poor resolution.) How much sample to apply in a spot or band is sometimes difficult to determine as it will depend on a number of variables, *e.g.* the sample matrix itself, the sorbent layer thickness, the nature of the sorbent, and the sample solvent. In most cases the following "rule-of-thumb" can be applied:

For HPTLC: 50–200 ng per spot, 1–4 μg per10 mm band
For TLC: 0.1–2 μg per spot, 2–10 μg per 10 mm band

As mentioned earlier, sample overloading is a common reason for poor chromatographic separations on thin layers. In order to understand clearly why sample overloading causes problems, one needs to understand more about the equilibrium ratio between the solute in the stationary and mobile phases. As development of the sample spot or band occurs, an equilibrium ratio establishes itself between the fraction of the solute in the stationary phase and that in the mobile phase. The ratio of the equilibrium concentration of the solute in the stationary and mobile phase is called the distribution coefficient (K), where K is:

$$K = \frac{c_s}{c_m} \tag{10}$$

where c_s is the solute concentration in the stationary phase
         c_m is the solute concentration in the mobile phase

Components of a mixture will separate if the value of K differs sufficiently under a given set of conditions. The greater the value of K, the greater the time that the solute resides in the stationary phase. Conversely, the lower the value of K, the greater the time the solute resides in the mobile phase. Therefore, during development, separation occurs as analytes with low K values are retained in the mobile phase and hence are transported further through the sorbent layer than those with high K values. Of course whilst in the stationary phase, solute molecules do not move at all.

For an ideal situation, the ratio of c_s to c_m will follow a linear correlation (see Figure 5a), resulting in the scanned peak A in Figure 5b. However if the spot/band is overloaded, the relationship of c_s to c_m becomes non-linear (see Figure 5c), and the scan in Figure 5d is the result. Notice also the positions of peak A in chromatograms 5b and 5d. Overloading of sample usually results in a slight increase in R_f value. This increases with loading concentration as shown in Figure 6. Obviously non-linear curves are undesirable, not just because they cause the R_f value to vary, but they can impair critical separations due to "tailing" effects and introduce unnecessary errors in quantification.

If streaking persists after optimising the loading, a methodical approach will be necessary to locate the problem. Usually it is due to unexpected interactions between the sample components and the sorbent. Ionic species in the sample will interact with acid or basic silanol groups in the sorbent layer causing unexpected

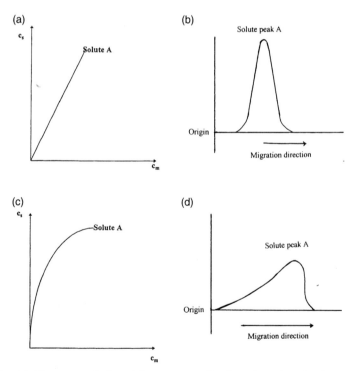

Figure 5 (a) *Linear correlation giving constant K values for solute A after satisfactory sample loading concentration;* (b) *Satisfactory loading of sample giving on development an ideal Gaussian peak;* (c) *Non-linear correlation for solute A as the sample loading concentration increases;* (d) *Overloaded sample giving on development a misshapen peak with a tail reaching back to the origin*

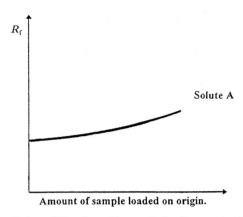

Figure 6   *Typical variation of R_f value with sample loading on the TLC plate*

retarding of analytes as the mobile phase migrates through the sorbent. (How to overcome these apparent separation problems is explained in Chapter 5.) On more rare occasions the fault may be the stationary or mobile phase. With the present precision and reproducibility of manufactured pre-coated TLC/HPTLC plates, diffusion should be kept to a minimum and few bands or spots will be misshapen. Hence, there should be few occasions when the sorbent layer can be blamed. Some mobile phase mixtures involving very polar and non-polar solvents are susceptible to solvent demixing in the vapour phase and give rise to a number of distortion problems during development, (this is described further in Chapter 5).

   The position of sample application is also important. All TLC/HPTLC plates will suffer from edge effects due to the method of manufacture of the layers; usually one edge will be slightly thicker than the rest. Slight distortions in chromatographic zones and variations in migration rates will be observed if samples are applied too near to the edge. For TLC and HPTLC plates it is advisable to apply samples at least 15 and 10 mm respectively from the edge for linear development.

4.1 Layer Damaging Effects

The way the sample is applied to the sorbent layer is also important. Layer damage can easily occur when pointed glass capillaries or syringes prick or scrape the surface of the sorbent layer as the sample is being applied. The softness of the layer contributes to the effect as do multiple applications to the same spot, particularly where insufficient drying has taken place between each addition. As a general rule, the more polar the solvent used, the more easily damage occurs. The result of such damage, like sample overloading, is not immediately apparent until development of the chromatogram is in progress. Usually peculiar shaped zones are observed with thin following streaks. If the original damage results in a hole in the layer, the sample concentrates in a ring around the hole. On development, the chromatographic zones often appear triangular or sometimes crescent shaped, (see Figure 7).

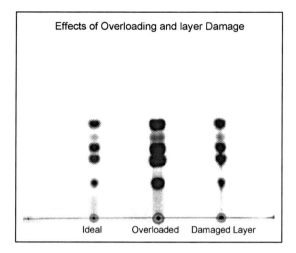

Figure 7 *Effects of incorrect loading of sample on the sorbent layer*

4.2 Humidity Considerations

Knowledge of the effect of humidity on migration of analytes can be a useful variable in the optimising of resolution. Moisture present in the layer affects the stationary phase and hence the partitioning of the analytes between the stationary and mobile phases. However, often where the effect of humidity is not taken into account, there can sometimes be quite unexpected results. Problems can be caused by the rate of evaporation of solvent from the applied sample, and incomplete drying of the applied sample spots or bands. As volatile solvents evaporate from the surface of the sorbent, the latent heat of evaporation required is taken from the surrounding atmosphere, causing localised cooling. With the cooling of the atmosphere close to the sample spot/band, moisture condenses on to the layer and is adsorbed. Unless the area of application is dried thoroughly at an elevated temperature, the migration of the components of the sample may proceed faster than expected. However, care must be taken to ensure that the temperature is not too high to cause degradation of the sample. It is therefore vitally important that the applied spots or bands are dried using an oven or even a hair dryer. Usually about 2–5 minutes at 70–80 °C should suffice for most applications.

5 Sample Application Methods

Application of the sample to the sorbent layer can be performed manually with very simple equipment or automated using sophisticated instrumental methods. As a general rule, the more precise the positioning of the sample delivery and delivery rate, the more reliable will be the final chromatographic results. Partially and fully automated application instruments are commercially available that can dramatically improve the quality of sample application. It should be remembered that the chromatographer is trying to achieve sharp regular spots or zones on the sorbent layer without disturbing that layer unduly. This section will suggest ways in which manual application can give satisfactory results for most qualitative separations, and will further show the advantages of instrumental band applicators to give improved resolution and reproducible quantification.

5.1 Manual Methods

Before the sample is applied to the plate, the origin point should be marked in some way so that migration distances can be determined after development. Often pencil lines are used, but these should be avoided if possible, as they can damage the layer. A better method is to make pencil marks at the edges of the plate. After development, the two marks can be joined by a pencil line to indicate the position of the origin. Spotting guides are a much better alternative as sample loadings are then made in a precise way at a predetermined distance from the edge of the TLC plate without any marking or disturbance of the sorbent layer (see Figure 8).

The simplest method of application of the sample solution to the TLC layer is by using a standard glass capillary of set volume. A capillary of 1–2 μl will give a spot size of approximately 3–4 mm. Calibrated disposable glass capillaries are widely available, and include "Microcaps" (produced by Drummond Scientific Co.) and the capillary dispenser system (from Camag). The latter have set capillary volumes of 0.5–5 μl with high reproducibility (>0.25%) (see examples in Figure 9).

Microsyringes allow a variable volume to be applied, but it is important to ensure that a flat-ended, preferably PTFE-coated needle is used to avoid disruption to the sorbent layer. With all these types of applicators, the capillaries or syringe needles must contact vertically and at right angles to the sorbent layer, and for quantitative work the whole contents must be administered at a controlled flow rate in one

Figure 8   *A multi-purpose spotting guide for a 20×20 cm TLC glass plate. One side of the pipette rest has 11 notches, resulting in sample loadings 15 mm apart. Scored lines can be drawn with the tapered pen through the long slit in the pipette rest* (Reproduced by permission of Camag, Muttenz, Switzerland)

Figure 9 *A Camag capillary dispenser unit. The glass capillaries are held in dust-proof dispenser magazines. Capillaries of 0.5, 1, 2, and 5 µl volume are available. To load the capillary holder with a capillary, the magazine is inserted into the dispenser, the holder is pushed into the slotted hole, and then pulled out. The capillary is now in position in the holder. After use, the capillary can be ejected by pressing the end of the holder*
(Reproduced by permission of Camag, Muttenz, Switzerland)

application. This will help to eliminate variations in spot shape and ensure spot sizes are consistent.

Band application with manual techniques is almost impossible without some damage to the TLC layer. It is also very difficult to attain a uniform concentration over the band length and width, and to obtain a straight line application. As described earlier, surface damage by capillaries is a major problem in spot application. For the best results, it is therefore important that a procedure is used whereby contact with the sorbent surface can be avoided if at all possible. Over the years special manual applicators for band dosage have been described in the literature, involving specially shaped capillaries with a reservoir of sample solution. However, although these can result in a good straight line application of sample, over most of the plate width, it is difficult, if not impossible, to obtain a uniform concentration over the whole band length and width.

5.2 Instrumental Methods

5.2.1 Manual Instrumental Techniques

For the aforementioned reasons, the best methods for sample application to the chromatographic layer are automatic or at least semi-automatic. In the case of spot

Figure 10   *The Camag Nanomat 4 with capillary holder in position. Sample spots can be applied precisely positioned without damage to the sorbent layer. The volume loaded is governed by the size of the disposable glass capillaries in the capillary holder. The unit can be used with all sizes of TLC and HPTLC plates*
(Reproduced by permission of Camag, Muttenz, Switzerland)

applications, the sample can be introduced to the layer surface at precisely the location desired, in one smooth dosage of the required volume, and with minimal or no damage to the layer. An example of such an instrument for fixed volume sample application to the TLC layer, is the Nanomat 4 with capillary holder, shown in Figure 10.

With this instrument, capillary sizes from 0.5–5 μl can be used and spots can be applied to all sizes of TLC plates. The capillary holder is held in position by a permanent magnet. To dispense the contents of the capillary, the applicator head is pressed down, the pipette touches the layer surface at a constant pressure, and the pipette is discharged. Variable volume dosage units have also been used with this device. The volumes dispensed are 0.5–2.3 μl and 50–230 nl for TLC and HPTLC respectively. The variable volume unit functions on the principle of a micrometer-controlled syringe with the height of the tip adjustable such that it does not touch the TLC layer (nominally a 50 μm gap). A lever is pressed to dispense the contents of the syringe on to the layer.

Also for spot application to HPTLC plates, fixed volume nano-pipettes (100 or 200 nl) have been employed. These have been made of a reagent-resistant platinum-iridium alloy fixed in a sealed glass holder.[4] This pipette allows accurate, reproducible loadings to the HPTLC layer by capillary attraction. However, the sorbent is susceptible to damage from the capillary particularly the softer HPTLC layers. Unless only light contact is made with the layer, the orifice of the nano-pipette can quickly become clogged with sorbent.

5.2.2 Automated Techniques

Automated equipment can be used to apply the sample solution as spots or bands. The equipment can vary from semi-automated to a complete robotic system controlled by a pre-set program on a computer. For accurate quantitative determination of separated analytes, such equipment is to be recommended. Although a degree of accuracy is possible with manual applications (approximately 1–2% relative standard deviation), automated spot and band loadings result in excellent reproducibility and reliable quantifiable separations. Automated band application equipment provides a very even, narrow application of sample as the sample syringe passes over a pre-determined length on the TLC plate at a constant speed, delivering a pre-set quantity of sample by a spray-on technique (see Figure 11). A constant flow of gas ensures that both a very thin application zone is possible and the band is dried quickly when volatile solvents are used. In this way the surface of the TLC layer never comes into contact with the syringe. Such an automated application of 1–20 μl results in an extremely thin line with little diffusion of the sample on either side. The equipment can best be described as semi-automated, as the syringe is cleaned and refilled with each sample solution manually.

For fully automated systems, programs of sample application can be stored on a personal computer. Both spot and band application can be programmed, with details on number, size and position of applications. An example of such a robotic system is shown in Figure 12. Bands are applied using a spray-on technique in a

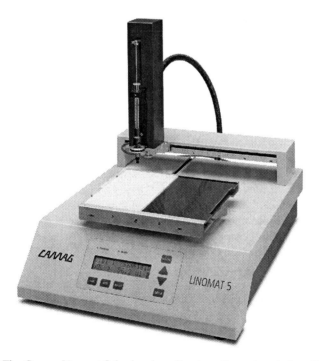

Figure 11 *The Camag Linomat 5 for band application of sample solution. This spray-on technique allows larger sample volumes at low concentration to be applied than is possible by contact transfer. Samples are concentrated into narrow bands of selectable length. For preparative purposes, sample solutions can be applied as long narrow bands over most of the plate length*
(Reproduced by permission of Camag, Muttenz, Switzerland)

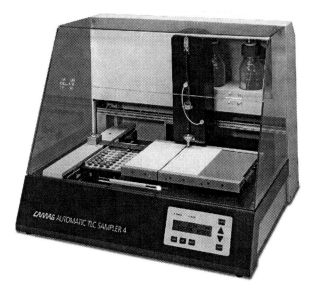

Figure 12   *The Camag Automatic TLC Sampler (ATS) 4 for spot and band application*
of sample solutions from a rack of vials. The operation is programmed
and controlled by a personal computer. A standard rack of 66 2 ml vials
may be addressed in one application program. Optional special racks hold
96-well plates. The recommended volume range per sample is 100–5000 nl for
spots and 2–20 µl for bands
(Reproduced by permission of Camag, Muttenz, Switzerland)

similar way to the semi-automated units. Spots can be applied either by this
technique or by direct contact transfer. The samples are prepared in vials with
septum seals. According to the pre-set program, the robotic arm will move from
vial to vial removing sample solution into a syringe and delivering it on to the
chromatographic layer at the pre-determined point. Between sample applications
the robotic arm will move the syringe to a bottle containing an appropriate wash
solvent. After several flushes the syringe is returned to the next sample vial. The
software allows the user to validate the instrument at appropriate intervals. Volume
dosage can be validated using a standardised method.

5.3 Large Sample Application

5.3.1 Contact Spotting[5]

In order to avoid damaging layers with capillaries and at the same time to apply large
volumes (~50 µl) as sample spots to the TLC layer, contact spotting has been used.
This technique also allows the application of highly viscous biological samples, not
easily applied by other devices. Spot sizes of 1 mm and less can be obtained.

The sample is placed in a depression or dimple in a fluorocarbon polymer film
that has been pre-treated with a coating of perfluorokerosene (or similar
perfluorinated fluid). A symmetrical droplet forms in the depression (the depression
is large enough to accommodate at least 50 µl of sample solution). With the

apparatus covered, a gentle stream of nitrogen is passed over the surface to evaporate the solvent. After complete evaporation, a TLC plate is placed sorbent side down over the sample positions and clamped. Nitrogen pressure is applied at approximately 1.5 atm. to transfer the samples to the sorbent layer.

5.3.2 Use of Concentration Zone TLC/HPTLC[6,7]

Commercially available concentration zone TLC/HPTLC plates described earlier in Chapter 2 are ideal for large sample volume applications. The sample is applied to the concentration zone area, usually as dilute large spots, in any location desired within the zone or by immersion over the complete width of the plate (Figure 13). Loadings of sample on the layer will vary depending on the resolution possible between separated components. For example with the separation of lipophilic dyes shown in Figure 1, loadings up to 20 μg were applied and good resolution still obtained, (20 μl loading of a 0.1% w/v solution using a 2 μl pipette).

If sample solutions are very dilute and concentration techniques are not an option, then repeated applications can be made directly to the concentration zone, with drying between loadings as required. When the plate is introduced to the mobile phase in the chromatography chamber, the solvent front migrates rapidly

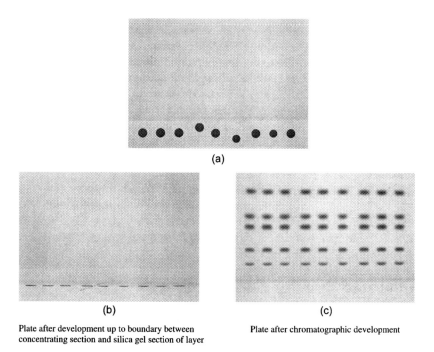

(a)

(b)   (c)

Plate after development up to boundary between   Plate after chromatographic development
concentrating section and silica gel section of layer

Figure 13 (a) *Concentration zone plate after application of sample spots* (b) *Plate after migration to the boundary between concentrating zone and silica gel separation layer* (c) *Plate after chromatographic development*
(By permission of Merck)

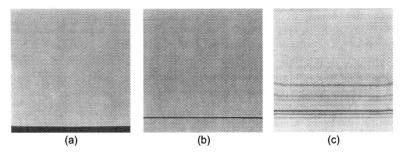

(a)	(b)	(c)

Figure 14   (a) *Plate with concentration zone impregnated to a height of 10* mm (b) *Plate after migration up to the boundary* (c) *Plate after chromatographic development* (By permission of Merck)

through the concentration zone carrying the samples with it. Ideally as the concentration zone material has only a low retentive capacity the sample spots begin to concentrate into bands. By the time they reach the interface they have become sharply focused. Some impurities are retained by adsorption in the lower zone and separation of the sample components then begins in the upper zone of silica gel 60. Resolution is better than for normal spot development, as sharp discrete bands are formed particularly for the most retained components that remain near the origin. However, the quality of focusing depends also on the choice of mobile phase. Generally for silica gel 60 low polarity, low viscosity solvents give the best results. Polar, viscous solvents as mobile phases can cause only minor improvement in separation. More detail on these advantages and the benefits of concentration zone TLC/HPTLC has been demonstrated by Halpaap and Krebs using a series of lipophilic dyes.[8] As shown in Figure 14, the highest possible loadings for a thin layer can be obtained by using the whole width of the plate and dipping this into the sample solution. After drying this can be repeated until the required loading is obtained, but care is needed to avoid overloading. Any leakage into the mobile phase reservoir is usually negligible.

To increase the potential of concentration zone layers, reversed-phase plates have now come into use and have proved valuable for a wide range of applications. Instead of silica gel 60, silica gel RP18 has become the separation medium with an appropriate wide pore sorbent for the concentration zone.

6 Choice of Band or Spot Application

Even for basic qualitative TLC, there are advantages in band over spot application. Although spots can easily be applied manually using a glass or metal capillary, the application of sample as a band usually requires more dexterity and is more accurately accomplished with semi- or fully automated equipment, unless concentrated zone TLC plates are incorporated. One of the major advantages of band application is illustrated in Figure 15 with a sample dye chromatogram with spot development compared against concentration zone band development. Although for most components good resolution is obtained for both using the

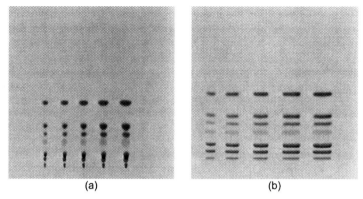

Figure 15 *Comparison of separation between spot and band application. Development after application as:* (a) *spots,* (b) *bands*
(By permission of Merck)

same developing solvent and conditions, it is in the region of low R_f, that major differences are noted and band development gives better results.

Because bands are formed as a thin line, whereas spots cover a wider area above and below the origin, the separation resolution in the low migration area is very sharp for bands in comparison with spots. Using the equation mentioned earlier and proposed by Kaiser[3], the spot capacity or separation number can be calculated from the following:

$$SN = \frac{Z_f}{b_0 + b_1} - 1 \tag{6}$$

Table 1 shows a comparison of results for spot and band separation based on this calculation from experimental results. For low loadings of sample, the separation numbers are almost equivalent, but as the concentration is increased the separation number dramatically drops for separation of spots compared with bands.

This illustrates one of the major advantages of band over spot application. There are a number of others as outlined in Table 2. Obviously when a spot is applied to the sorbent layer, even though a low polarity solvent may have been chosen, there will still be a concentration of the solute in the centre of the spot with a much lower concentration at the perimeter. After development, further diffusion into the layer will have occurred. For spectrodensitometric determination it will therefore be necessary to choose a slit length sufficient to cover the whole of the largest developed spot. For band application, a slit length of two-thirds of the band length is recommended. However, there is much more flexibility here than for spot scanning. As spots do not always migrate in a precisely vertical direction (for normal ascending development), automated track adjustment is required on the scanning equipment; this is not necessary for band scanning.

These advantages can result in greater accuracy. As a practical example, comparing spot application using a variable volume nano-pipette with a semi-automated band applicator, the relative standard deviation can drop from 1–2% to

Table 1 *Comparison of separation numbers (SN) for silica gel 60 pre-coated TLC plates with and without concentration zone. Chromatograms run with a mixture of 7 lipophilic dyes, 0.1% in toluene. N-chamber used without chamber saturation, mobile phase: toluene, $Z_f = 100$ mm*

Sample applied (μg)	Separation number for TLC plate without concentration zone	Separation number for concentration zone TLC plate	SN for concentration zone plate/SN for plate without concentration zone
0.01	11.22	12.83	1.14
0.10	11.20	12.82	1.14
1.00	11.02	12.75	1.14
2.00	10.82	12.68	1.17
3.00	10.63	12.61	1.19
4.00	10.44	12.54	1.20
5.00	10.25	12.46	1.22
10.00	9.37	12.11	1.29
20.00	7.82	11.43	1.46
30.00	6.53	10.79	1.65
40.00	5.45	10.19	1.87
50.00	4.55	9.62	2.11
100.00	1.85	7.22	3.90

(By permission of Merck)

0.5–1%. The advantages of spot application are mostly related to the simplicity of equipment, and above all the cost of the technique. As we know it is very easy and quick to apply a sample with a glass capillary. However, damage to the layer can result and capillaries can become clogged with sorbent. It really depends on the

Table 2 *Comparison of band and spot application of sample to the chromatographic layer*

Advantage of band over spot application	Advantage of spot over band application
Better resolution of analytes – particularly near origin, hence higher separation number	Can require less automation
More even distribution of sample	Can be very inexpensive
Less error on choice of scanner slit length (automation)	Application usually less time consuming
Greater accuracy – lower % standard deviation	
More flexibility in sample loadings	

quality of separation one desires. For quick identification purposes, the spotting technique with basic equipment is quite adequate. But where precision is required with a high degree of accuracy, automated band application is recommended. This is particularly important where the separated zones are very close together. Using concentration zone plates is not the answer to simple band application, except at a purely qualitative level as during sample band focusing most of the solute concentrates in the centre of the band. Hence sample distribution is uneven along the length of the band.

7 References

1. G. Guiochon and A.M. Siouffi, *J. Chromatogr.*, 1982, **245**, 1–20.
2. D.C. Fenimore in *Instrumental HPTLC*, W. Bertsch and R.E. Kaiser (eds), A. Hüthig Verlag, Heidelberg, Germany, 1980, 81–95.
3. R.E. Kaiser in *HPTLC High Performance Thin-Layer Chromatography*, A. Zlatkis and R.E. Kaiser (eds), Elsevier, Oxford, UK, 1977, 21–32.
4. R.E. Kaiser in *HPTLC High Performance Thin-Layer Chromatography*, A. Zlatkis and R.E. Kaiser (eds), Elsevier, Oxford, UK, 1977, 88–90.
5. D.C. Fenimore and C.J. Meyer, *J. Chromatogr.*, 1979, **186**, 555–561.
6. D.C. Abbott and J. Thompson, *Chem. Ind.*, 1965, 310.
7. A. Musgrave, *J. Chromatogr.*, 1969, **41**, 470.
8. H. Halpaap and K.F. Krebs, *J. Chromatogr.*, 1977, **142**, 823–853.

CHAPTER 5

Development Techniques

1 Introduction

Thin-layer chromatography can be defined as a differential migration process where sample components are retained to differing degrees in a thin layer of sorbent as a solvent or solvent mixture moves by capillary action through the layer. The retention of analytes will depend on the interactions that occur with the liquid phase on the surface and contained within the porous physical structure of the sorbent, or there may be a direct interaction with the molecular structure of the sorbent. Also retention will be affected by the nature of the moving solvent and any additives contained in it. The types of interactions involved may be electrostatic, hydrogen-bonding, ion-exchange, size exclusion, and other van der Waals' forces. Sorbents are chosen to be inert to any reaction with the solvent or solvent mixture used for development of the chromatogram (often described as the mobile phase, but strictly the term only applies under development conditions), the solute, or the "liquid" retained in the sorbent (called the stationary phase).

2 The Theory of Solvent Migration

As the environment in the development chamber can constantly change during chromatography due to charging conditions in the sorbent, the migration rate of the solvent front does not occur in a linear fashion. As a general rule the movement of the liquid front as it migrates through the inert sorbent by capillary action is denoted by the following equation:

$$(Z_f)^2 = kt \tag{1}$$

Where Z_f is the distance from the immersion line to the solvent front
       k is the flow constant (or velocity coefficient)
       t is the time elapsed from the start of development

As can be seen from the above equation, there is a direct relationship between the migration distance of the mobile phase and the time taken for development. However, practical observation in the early stages of development indicates that the solvent front velocity is not constant. The further the solvent migrates, the lower the speed of migration becomes. This can be expressed mathematically by the following equation:

$$V_f = \frac{k}{2Z_f} \qquad (2)$$

Where V_f is the solvent front velocity

Flow rates will also vary according to the physical characteristics of the solvents used in the mobile phase. Both the viscosity and the surface tension will obviously affect the migration rate and have been mathematically related to the velocity constant (k) by the following formula:

$$k = 2\frac{\gamma}{\eta} K_0 d_p \cos\theta \qquad (3)$$

Where K_0 is the specific permeability

d_p is the particle size of the sorbent

θ is the contact angle of the sorbent with the mobile phase

γ is the surface tension

η is the viscosity

The specific permeability (K_0) is a dimensionless function of the external porosity of the sorbent layer. The average K_0 value for both silica gel 60 TLC and HPTLC plates is 8×10^{-3}. Using equation (3), the velocity constant (k) can be calculated, but the result obtained will depend very much on the value of the contact angle (θ). On normal silica gel the value of θ is less than $10°$ for organic solvents. Hence the value of Cos θ is greater than 0.985. For practical purposes where total wetting of the layer occurs the value of Cos θ approximates to unity. The equation can therefore be simplified to:

$$k = 2\left(\frac{\gamma}{\eta}\right) K_0 d_p \qquad (4)$$

From this it becomes very evident that the velocity constant (k) is dependent to a large extent on the surface tension and the viscosity of the solvent. (Table 1 lists the values of k for a range of solvents used in planar chromatography.) Consequently solvents with high viscosities and surface tensions will migrate at much slower rates than those with low viscosities and surface tensions.[1] It is therefore advantageous to mix solvents with low k values with those with higher values to improve migration rate, but care will need to be exercised to ensure that the solvents are miscible and that the appropriate polarity is maintained. The migration rates given in Table 1 are an experimental approximation to indicate the variation in flow rate with the viscosity/surface tension ratio.

It is also evident from equation (4) that the velocity constant (k) will increase as the particle size of the sorbent increases. The solvent front migration rate is therefore noticeably faster for TLC (10–20 µm) as against HPTLC (5–6 µm). Figure 1 shows the direct relationship between particle size and velocity constant for a particular solvent.[2]

Table 1 *Velocity constant (k) of TLC solvents on silica gel 60 HPTLC plates compared with viscosity and surface tension data*

Solvent	Velocity constant $(k)\ cm^2\,s^{-1}$	Viscosity (cP)	Surface tension $N\ m^{-1}$
n-Hexane	0.118	0.31	18.42×10^{-3}
Acetonitrile	0.114	0.35	19.10×10^{-3}
Acetone	0.112	0.36	23.32×10^{-3}
Tetrahydrofuran	0.103	0.55	27.31×10^{-3}
n-Pentane	0.092	0.24	15.48×10^{-3} (25 °C)
Water	0.082	1.00	80.10×10^{-3}
Ethyl acetate	0.080	0.45	23.75×10^{-3}
Chloroform	0.079	0.57	27.16×10^{-3}
Toluene	0.071	0.59	28.52×10^{-3}
Methanol	0.050	0.59	22.55×10^{-3}
1,4-Dioxan	0.050	1.37	33.75×10^{-3}
Cyclohexane	0.047	0.94	24.98×10^{-3}
Ethanol	0.031	1.22	22.32×10^{-3}
Propan-2-ol	0.019	2.40	21.79×10^{-3} (15 °C)

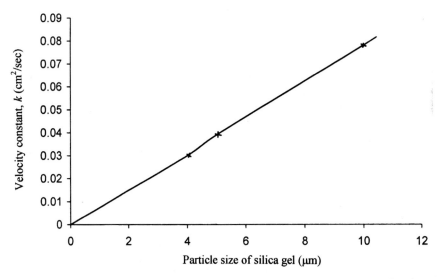

Figure 1 *Variation of the velocity constant (k) with particle size of silica gel 60 (solvent: dichloromethane)*

Table 2 *The variation of the cosine of the contact angle (θ) with solvent or solvent mixtures for RP18 reversed-phase plates*

Solvent	Cos θ
Toluene	0.96
n-Heptane	0.94
n-Pentane	0.87
Chlorobenzene	0.83
Acetonitrile	0.75
Methanol	0.84
Methanol/water (90:10)	0.59
Methanol/water (80:20)	0.35
Methanol/water (70:30)	0.24
Ethanol	0.87
Ethanol/water (90:10)	0.76
Ethanol/water (80:20)	0.61
Ethanol/water (70:30)	0.48
Ethanol/water (60:40)	0.34
Ethanol/water (50:50)	0.14

For reversed-phase TLC/HPTLC the value of the contact angle plays a much more significant role. Solvents and solvent mixtures used in reversed-phase TLC are quite polar, usually mixtures of water and methanol, ethanol, acetone, THF, and acetonitrile with various additives. Hence when these come into contact with a very non-polar silica gel layer bonded with a saturated carbon chain, wetting is slow. As the percentage of water is increased in the mobile phase, the wetting angle (θ) becomes larger, leading to a corresponding decrease in the velocity constant (k) until a point is reached where the mobile phase no longer migrates through the sorbent. At this point, the velocity constant (k) is infinitely small. Table 2 lists the comparison of solvent with Cos θ and demonstrates how the value of Cos θ drops with an increase in water content. In order to overcome this solvent "wetting" problem, many commercially available bonded reversed-phase TLC and HPTLC layers are only partially silanised. The total carbon loading is consequently lower, but the separation mechanism that predominates is still reversed-phase partition chromatography. Such bonded layers will withstand high aqueous content solvent mixtures as mobile phases. In fact, in many instances even total aqueous phases can be used although it would be debatable whether much genuine partitioning of the sample analytes occurs during development.

3 Mechanisms of TLC Separation

3.1 Introduction

There are three main types of separation mechanism in TLC:

Adsorption
Partition
Ion-exchange

Quite often the separation that occurs cannot be attributed to just one mechanism, but may be the result of two or more different types of interaction including those above. Other interactions are also known and may at times play an important part in the separation mechanism. Included in these are ion-pairing, charge-transfer, and π–π interactions. Usually a deliberate attempt is made to use these interactions to change the selectivity to improve resolution, as for example in the separation of enantiomers, complex mixtures of polycyclic aromatic hydrocarbons, or *cis/trans* isomers of lipids or fatty acids. These complex types of interaction will be described later in this chapter under the appropriate headings.

3.2 Adsorption Separations

Adsorption is a surface phenomenon. In the case of the sorbent silica gel, the sample interacts with the silanol groups on the surface (as shown in Figure 2). Dipole-induced interactions occur depending on the nature of the solute. As the mobile phase migrates through the sorbent by capillary flow, weakly adsorbed analytes move with the solvent front, whereas those more strongly bound are retained nearer the point of application of the sample. Hence separation occurs based on the strength of the dipole-induced interaction.

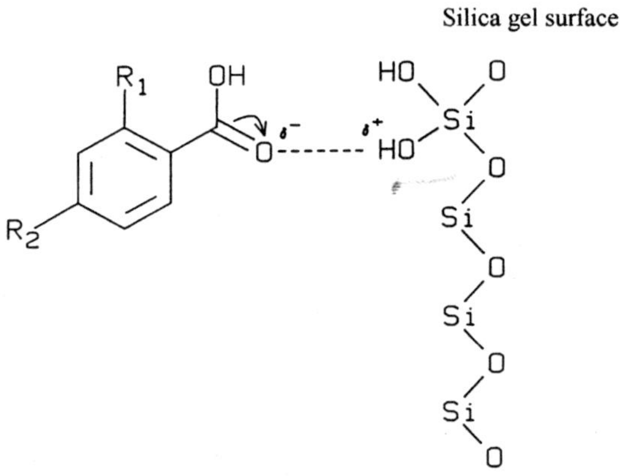

Figure 2 *Retention of analytes by silanol interaction on the sorbent surface. Strength of the retention is dependent on the strength of the induced dipole*

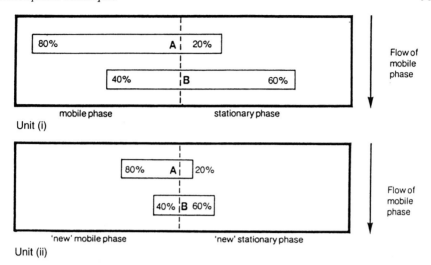

Figure 3 *Unit* (i) *is the first portion of stationary phase under the effect of flowing mobile phase where separation can begin. Unit* (ii) *is the second portion of stationary phase. The height of these portions is that required for equilibrium*

3.3 Partition Separations

In partition TLC,[3,4] a liquid stationary phase is adsorbed or chemically bound to the silica gel support. The stationary phase should ideally be insoluble in the mobile phase. On application of the sample to the sorbent layer, the components of the sample equilibrate between the mobile and stationary phases. The migration of components is dependent upon their relative solubility or preference for the mobile or stationary phases. On this basis, a component that has greater affinity for the mobile phase will be eluted before one that has more affinity for the stationary phase. To illustrate how this works consider a small portion of solid support containing a unit of stationary phase, (see Figure 3, unit (i)). A mixture of two substances, (A) and (B) is applied to the solid support. As the mobile phase flows through this unit, equilibration occurs and 80% of (A) and 40% of (B) prefers the mobile phase. As the mobile phase passes on to the next unit as shown in Figure 3, unit (ii), the sample remaining in the stationary phase will re-equilibrate itself. This process will continue with further equilibrium stages and substance (A) will migrate faster than substance (B). On a TLC or HPTLC plate there are many such equilibrium stages that are known as theoretical plates. Resolution obviously improves with the number of theoretical plates. As a general rule for TLC, the number of theoretical plates is >600 and for HPTLC typically >5000.

3.4 Ion-exchange Separations

The ion-exchange process is dependent on the sorbent containing ions that are capable of exchanging with ions of like charge in the sample or mobile phase. The exchange of ions is dependent on the affinity of the support for the various ionic species. The mobile phase acts as an electrolyte solution. The migration rate of the components of the sample is pH dependent as each component has a total charge, which changes with increased or reduced acidity of the mobile phase. Hence at constant pH, compounds with a low affinity exchange easily with mobile phase ions and migrate with the solvent front, whilst compounds with a strong affinity migrate slowly, and remain near the origin.

4 Solvent Selection

Selectivity of separation is greatly influenced by the choice of solvent or solvent mixture. This is a critical area of the separation technique in TLC and particular care and diligence here can make all the difference between a well resolved chomatogram and a poor diffuse streak in the direction of migration of the mobile phase. Often incomplete resolution, high or low retention, distorted analyte spots or bands and tailing can be blamed on the solvent used. As a general rule, non-polar solvents will effect migration of low polarity substances, whilst more polar samples will require more polar solvents on a normal- phase sorbent layer. This should come as no surprise as both mobile phase and sample analytes will compete for the adsorbent sites on the stationary phase.

Every effort should be made to simplify the components of the mobile phase as much as possible. If possible, a single solvent is to be preferred, as the more complex the mixture, the more problems of solvent "demixing" and vapour phase saturation can occur. All solvents should be of chromatographic grade purity with particular attention paid to low impurities, moisture content and non-volatile matter. Solvent mixtures should be thoroughly shaken together in order to attain complete homogeneity. These should in most cases, be used immediately after preparation as some mixtures can chemically react together resulting in an unwanted species in the final eluent. Fortunately this is usually a slow process where it does occur, and does not normally cause any appreciable difference in resolution in the short term. Care is also needed with solvent mixtures where partial immiscibility occurs. In these cases, either the upper or lower layer is used as the mobile phase, but it is important that the two layers are well separated in an appropriate separating vessel and the layer to be used is carefully specified.

In the past, most of what is known about solvent selection has come about by trial and error experiments by experienced thin-layer chromatographers. Usually when silica gel 60 is chosen as the sorbent, an eluent of low solvent strength, *e.g.* hexane, benzene, toluene, cyclohexane or dichloromethane is used for development. If little or no migration occurs, progressive additions are made of solvent or mixtures of solvents of higher strength until good resolution is obtained with R_f values ideally in the range 0.2–0.5. With reversed-phase separations the solvent starting point is far more polar, usually acetonitrile or methanol mixed with water. However, it is true to say that in recent years a more fundamental mathematical approach has been adopted. Also much more is now understood about the way in which solvents migrate through the stationary phase, the role of the vapour phase, and the problems of multiple-front formation. These are some of the areas that will be considered later.

4.1 Eluotropic Series and Solvent Optimisation

In order to classify solvents and mixtures of solvents into a solvent strength order, thin-layer chromatographers have devised tables to make the process of solvent selection easy. Some have used water solubility, and dielectric constant as their criteria.[5,6] Solubility data although useful, is limited and by no means gives the complete picture of solvent strength. Dielectric constants give a much better alternative, but even here there are abnormalities, *e.g.* 1,4-dioxan and tetrahydro-furan have values of 2.2 and 7.4, respectively, but are known to exhibit high polarity with silica gel compared with dichloromethane, even though it has a dielectric constant of 8.9. It has therefore been necessary to consider a much more complete alternative where both the characteristics of the solvent and stationary phase are taken into consideration. Several models have been proposed, the most universally accepted being that according to Snyder.[7,8] A solvent strength parameter depending not only on the appropriate physical characteristic of the liquid but also on the sorbent can be defined. This is called ε^0 and is defined as the adsorption energy per unit of standard solvent for a given solvent/sorbent combination. Values can be calculated for a wide range of solvents used in TLC and a table constructed called the eluotropic series in order of ε^0 values. Alumina was originally the sorbent of choice, but some values for silica gel can also be obtained by experiment as shown in Table 3. However, theoretical values for silica gel can be estimated from the following equation:

$$\varepsilon^0_{silica\ gel} = 0.77 \times \varepsilon^0_{alumina} \tag{5}$$

As seen from Table 3 there are only minor discrepancies between $\varepsilon^0_{expt.}$ and $\varepsilon^0_{calc.}$ values for silica gel. ε^0 is independent of the activity of the sorbent, and by definition, is equal to zero for n-pentane.

If the resolution is either poor or not possible with a single solvent, then binary or ternary mixtures can be prepared. This greatly expands the number of solvent systems and allows intermediate values of ε^0 to be calculated. For cases where the migration rate of sample components is well optimised, adjustments can be made with other solvents to improve resolution, but always ensuring that the ε^0 value is maintained. To help in these cases an approximate relationship has been derived by Snyder[9–11] to calculate the intermediate ε^0 values for binary mixtures, solvents A and B:

$$\varepsilon_{ab} = \varepsilon_b + \frac{Log\,N_b}{\alpha\,n_b} \tag{6}$$

ε_{ab} is the intermediate ε^0 value
ε_b is the ε^0 value for solvent B
n_b is the molecular area of adsorbed solvent B
α is the adsorbent surface activity function
   (for TLC silica gel at 40% RH, $\alpha = 0.69$)
N_b is the mole fraction of B in the solution phase

Table 3 *Eluotropic series for common solvents used in TLC for aluminium oxide and silica gel. Based on the solvent strength (ε^0) according to Snyder[7,8]*

Solvent	Alumina ε^0 value (expt.)	Silica gel ε^0 value (expt.)	Silica gel ε^0 value (calc.)
n-Pentane	0.00	0.00	0.00
n-Hexane	0.01		0.01
iso-Octane	0.01		0.01
Cyclohexane	0.04		0.03
Cyclopentane	0.05		0.04
Carbon tetrachloride	0.18	0.11	0.14
iso-Propyl ether	0.28		0.22
Toluene	0.29		0.22
Chlorobenzene	0.30		0.23
Benzene	0.32	0.25	0.25
Chloroform	0.40	0.26	0.31
Dichloromethane	0.42	0.32	0.32
Methyl *iso*-butyl ketone	0.43		0.33
Tetrahydrofuran	0.45		0.35
Diethyl ether	0.46	0.38	0.38
1,2-Dichloroethane	0.49		0.38
Acetone	0.56	0.47	0.43
1,4-Dioxan	0.56	0.49	0.43
Ethyl acetate	0.58	0.38	0.45
Methyl acetate	0.60		0.46
Amyl alcohol	0.61		0.47
Aniline	0.62		0.48
Acetonitrile	0.65	0.50	0.50
Pyridine	0.71		0.55
2-Butoxyethanol	0.74		0.57
Propan-1-ol	0.82		0.63
Propan-2-ol	0.82		0.63
Ethanol	0.88		0.68
Methanol	0.95		0.73
Ethanediol	1.11		0.85
Acetic acid	< 1		< 1
Water	≫ 1		≫ 1

Table 4 *Solvent strength table devised by Snyder et al. for reversed-phase liquid chromatography. Solvent strengths for mixtures of solvents can be calculated from this table using Equation 7*

Solvent	Solvent strength (S)
Water	0
Methanol	3.0
Acetonitrile	3.1
Acetone	3.4
1,4-Dioxan	3.5
Ethanol	3.6
Propan-2-ol	4.2
Tetrahydrofuran	4.4

In 1979 Snyder *et al.*[12] suggested a solvent strength table in reversed-phase liquid chromatography that could also be applied to TLC (Table 4). Although the values given are only approximate, they do provide a means of calculating a relative value for the solvent strengths of mixtures of solvents used in reversed-phase TLC. As the solvent strength for mixtures can be calculated from an arithmetic progression, it is a simple matter to determine the value for any volume mixture as long as S (solvent strength) is known for the pure solvent:

$$S_{mix.} = V_A S_A + V_B S_B + \dots \tag{7}$$

As in column liquid chromatography, mixtures of water, acetonitrile and methanol with sometimes polarity adjustments with THF, 1,4-dioxan or propan-2-ol are used mainly in reversed-phase TLC.

In normal-phase TLC the separation process is somewhat less predictable and hence it is more difficult to devise exact and reliable methods based on solvent strength values. The approach in such cases is much more general. An eluent polarity is selected on the basis of the sample nature and sorbent type. For silica gel, Table 3 can be used in conjunction with Equation (6) to establish the mobile phase required. Generally a solvent is chosen which gives an R_f value of between 0.2–0.5 (the region of maximum resolution efficiency on the layer). Sometimes the term hR_f is used, where:

$$hR_f = R_f \cdot 100 \tag{8}$$

If the R_f value approaches unity or the hR_f value 100, resolution will be poor and a solvent of lower ε^0 must be chosen (high solvent strength) in order to increase the retention of the analytes on the layer. The reverse would be true for analytes near the origin where a solvent with higher ε^0 would be required. Sometimes the term

R_m is used as it often shows a linear relationship between the chromatographic properties and the sample. R_m, the logarithm of the capacity factor (k') is related to the R_f value by the following equation:

$$R_m = \log k' = \log \frac{(1 - R_f)}{R_f} \qquad (9)$$

Once the desired migration has been achieved, the selectivity can be improved by mixing two or three solvents in such a ratio, as to maintain a similar ε^0 value. This may well take several attempts as there will no doubt be a number of combinations that are possible. However, the final result will be the selection of a mobile phase that gives the most favourable interactions with the solute. Often the best results are obtained by adjusting the main solvent of low polarity with a relatively small amount of solvent with a much higher polarity to attain the required ε^0 value, *e.g.* butan-1-ol with water, chloroform with methanol, dichloromethane with acetonitrile, and cyclohexane with dichloromethane. Further small adjustments with a third solvent can often fine tune the separation further.

Of course, the choice of mobile phase will be largely determined by the sample as mentioned earlier. With non-polar samples of hydrocarbons for example, low solvent strength (low polarity) mobile phases are required and progressively higher strengths (higher polarity) where more functional groups are present in the sample. The addition of small percentages of acids (*e.g.* formic, acetic, propionic and hydrochloric acids) or bases (*e.g.* ammonia solution, pyridine and amines) to the mobile phase can often improve resolution by suppressing ionisation or totally ionising the analytes. On the TLC layer, the spots or bands exhibit less diffusion and any hint of streaking is often eliminated, (see later section on mobile phase additives). A schematic for the optimisation of the mobile phase is shown in Figure 4.

This incorporates the above criteria and the classification of solvents by Snyder and Glajch[9-11] into eight groups according to their proton-donor, proton-acceptor or dipole interaction characteristics. This classification is shown in Table 5. Often solvent optimisation is overlooked as an unnecessary and tedious time-consuming procedure. However, it can make all the difference between well-resolved bands or spots and those that tend to merge together.

The above procedure is usually adequate for most optimisations of mobile phase for TLC. However, there are more mathematically based procedures often used in column liquid chromatography to optimise solvent mixtures. Two of the major ones used in TLC are the selectivity triangle and the PRISMA method.

4.1.1 Selectivity Triangles

The basis of these procedures is the selectivity triangle proposed by Snyder[13] and mentioned earlier. Solvents were classified into groups according to their similar proton-acceptor (x_e), proton-donor (x_d), and dipole interaction (x_n) contributions. It was possible to divide up the majority of solvents into eight groups as listed in Table 5. From this, Snyder was able to construct a selectivity triangle similar to that in Figure 5 with each corner of the triangle representing a single contribution.

Schematic for optimization of mobile phase

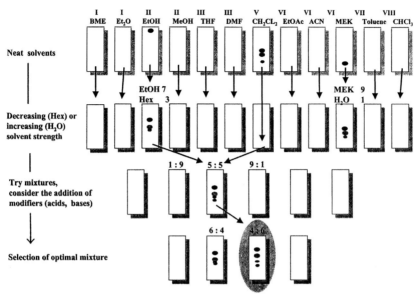

Figure 4 *Selection and optimisation of mobile phase.*
Stage (i): pure solvents; stage (ii): decreasing or increasing solvent strength by mixing two solvents; stage (iii): more complex mixing of solvents and addition of acid or base modifiers; stage (iv): selection of optimal solvent mixture
*BME is butylmethyl ether, THF is tetrahydrofuran, DMF is dimethylformamide, DCM is dichloromethane, EtOAc is ethyl acetate, ACN is acetonitrile, MEK is butanone, and CHCl₃ is chloroform

A further selectivity triangle can now be constructed using three solvents selected from separate groups that are near an apex of the original triangle. For normal-phase separations an example would be diethyl ether, chloroform, and dichloromethane. n-Hexane is added as a non-polar diluent to adjust the strength of

Table 5 *Classification of solvent selectivity according to Snyder and Glajch*[9-11]

Group	Solvents
I	Aliphatic ethers
II	Aliphatic alcohols
III	Pyridine, tetrahydrofuran, glycol ethers, amides (except formamide)
IV	Formamide, acetic acid, glycols
V	Dichloromethane, 1,2-dichloroethane
VI	Aliphatic ketones and esters, 1,4-dioxan, acetonitrile
VII	Aromatic hydrocarbons and ethers, aromatic halo- and nitro-compounds
VIII	Chloroform, water, nitromethane, m-cresol

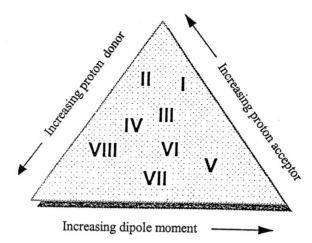

Figure 5   *Selectivity triangle according to Snyder. Lists of solvents for each solvent group are given in Table 6*

the developing solvent. For reversed-phase separations, acetonitrile, methanol, and tetrahydrofuran mixed with water form the corners of the selectivity triangle. As shown in the example in Figure 6 a further four mixtures are obtained at the mid-points of the sides of the triangle and at the centre. In all, seven mixtures are obtained. These can be run with a special developing chamber that allows applied samples to be run on separate discrete tracks on the same TLC/HPTLC plate, (see the Vario Chamber described later in this chapter).

The two best systems are usually then chosen. From a series of experiments using mixtures between these selected eluents, the R_f results can be extrapolated

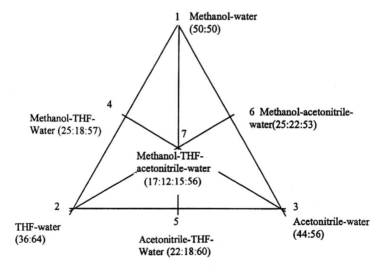

Figure 6   *Solvent selectivity triangle for optimising the mobile phase for reversed-phase separations*

to give the optimum mobile phase. However, it should be noted that the procedure does not take into account other chromatographic interactions that may have an additional influence on the separation.

4.1.2 PRISMA Model

A further development to this selectivity triangle is a prism. Nyiredy *et al.*[14-16] proposed such a system, calling it the PRISMA model. This is a three-dimensional model used to correlate the selectivity of the solvent with the solvent strength. Silica gel is taken as the stationary phase and a preliminary solvent selection made according to the previously described Synder classification using three solvents from the eight solvent groups. This is the first step in the optimisation. Step two involves the use of an unusual prism model, the upper part of which is an irregular frustum, and the middle and lower parts are regular triangular prisms as shown in Figure 7. The base of the prism represents the modifier (in the case of normal-phase separations this will be n-hexane with an ε^0 value of 0.00). The heights of the prism at each edge (S1, S2, and S3) represents the solvent strength for the neat solvents selected from step one. An isoeluotropic plane (where ε^0 values are equivalent) is formed by increasing the solvent strength at the corners of the prism. Points along the edges are therefore combinations of two solvents, points on the sides, three solvent mixtures, and points within the prism, combinations of four solvents. A modifier can be added to each solvent at the start, increasing the overall solvent mixture to five, if required.

In practice, for non-polar samples the starting point is the centre of the triangle on the top face of the main prism (marked A in Figure 7). Having run the

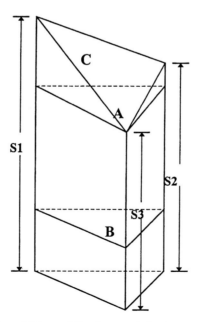

Figure 7 *The PRISMA model for mobile phase optimisation*

chromatogram under these conditions, the solvent composition is diluted with hexane until the solutes are in the required R_f range. The apexes of a triangle drawn through this plane (B) in the prism give three further solvent compositions that are now tried. These represent the extremes of selectivity for the solvent system. With the information gleaned from these initial chromatograms, selectivity points for further chromatograms can be chosen until the optimum solvent composition has been found. If the best chromatogram does not adequately resolve the analytes, then one or more of the primary solvents will need to be replaced and the optimisation procedure repeated. For polar solutes, the face C of the upper frustum is used for normal-phase chromatography and the solvent optimised in a similar way.

5 Development Chambers

There are a variety of different types of TLC chamber, each designed with particular features to control to a greater or lesser extent the parameters of chromatogram development reproducibility. As solvent vapour saturation, sorbent vapour adsorbed, solvent vapour "demixing", and solvent front and edge effects on the chromatographic layer can have a bearing on the separation achieved, it is important to eliminate unwanted effects and to utilise those features that will improve resolution. The following is a list of TLC/HPTLC chambers that will be described in detail:

Nu–chamber (normal flat-bottomed, vapour unsaturated glass tank)
Ns–chamber (normal flat-bottomed, vapour saturated glass tank)
Twin-trough chamber (two compartment tank, saturated or unsaturated)
Su–chamber (sandwich tank, unsaturated)
Ss–chamber (sandwich tank, saturated)
Horizontal chamber (twin development, saturated or unsaturated)
U–chamber (circular HPTLC, saturated or unsaturated)
Automatic development chamber (ADC) (fully environmentally controlled unit)
Forced flow development chamber (OPLC) (TLC development under pressure)
Vario chambers (saturated or unsaturated development using six different mobile phases on one sorbent layer)

5.1 N–Chambers and Chamber Saturation Effects

Chamber saturation means the point where all components of the solvent are in equilibrium with the vapour space before and whilst development of the chromatogram is proceeding.

Chromatographic development in TLC/HPTLC is really a process that occurs in two phases, the liquid phase and the gas or vapour phase at the same time. Prior to entering the development chamber, the sorbent layer will have adsorbed moisture from the atmosphere. Once the chromatographic plate is placed in the development tank, the layer immediately comes into contact with the vapour above the level of the solvent. A pre-loading of the sorbent layer with gaseous solvent molecules

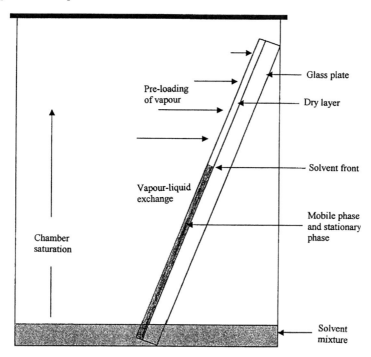

Glass plate

Pre-loading
of vapour

Dry layer

Solvent front

Vapour-liquid
exchange

Mobile phase
and stationary
phase

Chamber
saturation

Solvent
mixture

Figure 8 *Solvent and vapour phase effects in a standard N-chamber*

occurs, depending on the degree of saturation within the chamber. At the same time the solvent liquid begins to migrate through the capillaries in the sorbent and simultaneously interacts with the vapour phase (see Figure 8). There are therefore a number of interactions that can occur between the vapour phase, the solvent, moisture retained in the sorbent, and the sorbent itself. With the solute involved even more interactions are possible.

In the Nu-chamber or unsaturated N-chamber (Figure 8), 3–5 mm depth of solvent/solvent mixture is placed in the flat bottom and the tank is covered with a lid. A vapour phase gradient forms in the vapour phase space, and as time progresses the lower third of the tank becomes fully saturated (about 15 minutes), the middle section about 75% saturated, and the upper section less so. When the TLC plate is placed in the chamber, the whole gradient is temporarily upset and it takes some time for it to adjust to the new conditions. Chromatographic development now begins with the rising flow of solvent molecules in the sorbent layer. In the middle and perhaps in some of the upper part of the tank pre-adsorption of the vapour phase molecules occurs on to the dry layer. As the solvent front rises, a point is reached in the upper section of the tank where the vapour phase is no longer saturated. Evaporation then starts to occur from the solvent front into the vapour phase space and the velocity of the solvent front diminishes (see Figure 9). Although sample components separating with low R_f will be largely unaffected by this, components near the solvent front will begin to focus from spots into tight bands.

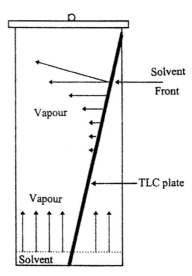

Figure 9 *Solvent vapour concentration effects in an unsaturated N-chamber. Saturation of the tank with solvent vapour occurs from the solvent mixture in the base of the chamber and from the solvents migrating through the sorbent layer*

Where solvents have low boiling points and are very volatile and of similar polarity, the above unusual effect is the only one observed. However, for low volatility situations, not only will this variation in solvent vapour be evident from top to bottom but also from the sides into the middle of the tank. This will be usually observed as concave solvent fronts and edge effects. Problems of this type can be easily overcome in a number of ways the most simple of which is by using a saturated chamber (Ns-chamber). In the Ns-chamber, the internal sides and ends of the tank are lined with adsorbent paper. Saturation in such a tank can be achieved in 5 minutes with very volatile solvents. In most cases it takes no longer than 15 minutes, particularly if the lining paper is pre-soaked in the developing solvent. When the TLC/HPTLC plate is placed in the tank, pre-loading of the dry layer occurs almost completely, within a few minutes, and the effects observed at the solvent front in the upper section of an unsaturated chamber do not occur.

Both saturated and unsaturated TLC tanks are used for a wide range of TLC separations. However, for a particular separation, it is important that the type of development used is recorded since R_f values for the same components resolved using the same mobile phase in both tanks will be different. Values will always be lower for saturated tanks due to multiple front formation.

For a saturated TLC chamber, it appears that the solvent migration through the sorbent layer is responsible for the observed solvent front. If that is truly the case, then a solvent front representing this effect can be called the "true" or "real" solvent front, sometimes called the virtual solvent front. However, ahead of such a "real" solvent front, condensation of solvent vapour from within the chamber occurs creating another solvent front. This is the observed or visible solvent front. Obviously, the more saturated the chamber is with solvent vapour, the more the dry

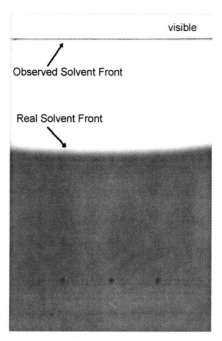

Figure 10 *Detection of the observed and "real" solvent fronts. The dye, fat red 7B migrates to the "real" solvent front. Developing solvent is dichloromethane. The tank is a fully saturated N-chamber*

layer ahead of the "real" front becomes pre-loaded with gas molecules. So, what is actually seen as the solvent front is a combination of solvent migration and condensation of solvent vapour.

It is possible to clearly demonstrate the position of the "real" front by a simple experiment (see Figure 10). A solvent-soluble dye (*e.g.* fat red 7B or sudan III) is added to the developing solvent before it comes into contact with the TLC plate in the chromatography chamber. Once migration of the solvent through the TLC layer occurs in a partially or fully vapour saturated chamber, two fronts are observed. The dye migrates with the "real" front and stains the whole layer below this line, but the section beyond and up to the observed front is wetted only by the solvent. Of course, if it is possible to run the chromatogram under conditions where no pre-loading of solvent vapour molecules occurs, then only one front is observed, the virtual or true solvent front (such chambers are described later).

Where mixtures of solvents are used, solvent "demixing" can sometimes occur. This usually is a result of the solvent mixture being composed of both very weak and high strength solvents. The presence of moisture in solvents or solvent mixtures can also in rare cases give rise to a similar effect. The result is the formation of what are called α- and β-fronts (see diagram in Figure 11).

In normal-phase TLC, the sorbent layer has initially more preferential adsorption for the more polar component rather than the less polar constituents. In order to

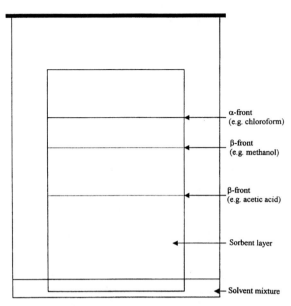

Figure 11   *The formation of α-and β-fronts during chromatographic development. The component of highest polarity in the mobile phase is adsorbed first in the sorbent, e.g. acetic acid (first β-front) followed by the others in order of decreasing polarity. Such solvent demixing on the stationary phase can occur either because of poor saturated vapour control, or due to too sharp a contrast of polarities of solvents in the mobile phase*

understand the demixing processes, consider a solvent mixture composed of n-hexane, ethanol, and acetic acid. As the development proceeds, the most polar component, the acetic acid is adsorbed forming an initially unobserved front, called a ß-front. The lesser polar component, ethanol has more mobility and forms another ß-front, again initially unobserved, but ahead of the first one. The least polar component, n-hexane migrates through the layer to the observed solvent front, called the α-front. Unfortunately it is only when the development is complete and detection reagents have been applied that the presence of ß-fronts becomes apparent. This solvent demixing effect on the sorbent layer often interferes with the migration rates of sample components resulting in misshapen spots or bands being restricted in migration to below or at ß-fronts. The same component may even migrate at different rates across the plate. Fortunately this effect rarely occurs as solvent "demixing" is not usually localised. Usually solvent "demixing" can be averted by fully saturating the TLC chamber with solvent vapour before chromatography, but with the dry layer present in the tank. In these instances the use of a standard N- or S-chamber is not possible and better results can be achieved with a twin-trough chamber. Attention may also be necessary in the choice of solvents for the mobile phase. Very polar solvents mixed with almost non-polar ones will sometimes result in "demixing". In such cases, using more of a slightly less polar solvent usually solves the problem.

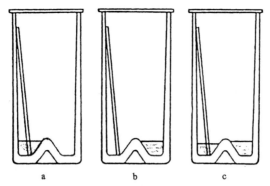

a   b   c

Figure 12 *Twin-trough chamber allowing solvent vapour pre-loading or humidity control.*
(a) *Standard N-chamber development with low solvent consumption (20 ml sufficient)*
(b) *Reproducible pre-equilibration with solvent vapour*
(c) *Vapour phase conditioning with any solvent or conditioning liquid*

5.2 Twin-trough Chamber

This is a useful, although simple modification to the N-chamber (see Figure 12).

It offers better control over the development conditions. It also has a number of other advantages. As Figure 12 shows, no more than 20 ml of solvent should ever be used per development (it can be as little as 4 ml!). If the solvent is carefully poured into one compartment, the plate can be allowed to pre-load with solvent vapour in the other (Figure 12b). With the 20×20 cm and 10×20 cm tanks there is not even the need to disturb the vapour concentration by removing the lid. After equilibrium, the tank can be tilted with the lid still in place to allow solvent to flow over from one compartment to the other starting the elution. Of course, the use of a second compartment also means that plates can be conditioned with other vapour phases before or during development. Sorbent layers can also be conditioned at required levels of relative humidity by introducing sulphuric acid or saturated metallic salt solutions into the second compartment or trough. This is sometimes an important parameter to control as some separations are susceptible to changes in relative humidity.

5.3 Sandwich, Horizontal, and U-chambers

Like the N-chamber, the sandwich or S-chamber can be used unsaturated or saturated (see Figure 13); however, most of the time it is used unsaturated. The purpose of the S-chamber is to produce a very thin tank restricting space for vapour phase saturation. At the beginning of development, the tank is therefore virtually unsaturated. During development the tank structure makes it difficult for vapour to enter and rise, so that little or no pre-loading of the dry sorbent layer ahead of the rising front occurs.

Horizontal chambers although in appearance quite different, are in fact very similar in principle to the sandwich chamber. How they function is shown in

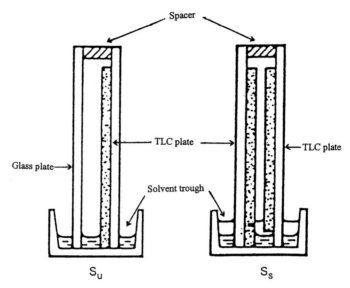

Figure 13   *Sandwich chambers. The S_u and S_s tanks are the unsaturated and vapour*
saturated chambers respectively
(Reprinted from F. Geiss in *Fundamentals of Thin Layer Chromatography,*
1987, p 301, Hüthig, Heidelberg, Germany with permission of the publisher,
Wiley-VCH and the author)

Figure 14. Only a few millilitres of developing solvent are required in the two
troughs positioned on opposite sides of the chamber. Samples can be developed
from one edge of the TLC/HPTLC plate or from both sides simultaneously.

The U-chamber, although a unit designed to run circular chromatograms, uses
the same principles of separation as the sandwich and horizontal chambers. It was
developed to exploit the improvements in resolution that can be achieved by the
fully controlled development of circular HPTLC chromatograms. Figure 15 shows
the principal parts of the chamber. A pre-spotted HPTLC plate (1) is placed with
sorbent layer downwards on the U-chamber body (2). Developing solvent is pumped
at a pre-determined rate onto the centre of the plate via a fine bore platinum–
iridium capillary (3). The atmosphere of the chamber can be altered at will by
passing the vapour phase desired through channel (4) and out through channel (5).
The R_f values measured differ from those measured in linear TLC. The two are
related by the following formula:

$$RR_f^2 = R_f \tag{10}$$

It follows from this, that circular chromatography is particularly useful for sample
components that show little migration from the origin. For example an analyte
which could have an R_f value of 0.01 (almost not measurable) would have an
RR_f value of 0.1 even though the mobile phase was the same. Also, although
the samples are applied as spots, on development arcs, or bands form for each

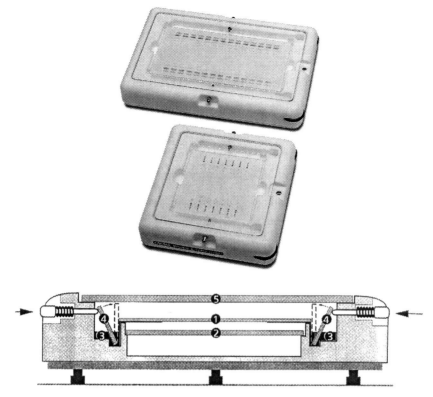

Figure 14 *The Camag horizontal developing chamber.*
The principle of operation of the Camag Horizontal Development Chamber in the sandwich configuration with development from both sides is described as follows: the HPTLC plate (1) is placed layer-down at a distance of 0.5 mm from the counterplate (2). If the tank configuration is to be used, the counterplate (2) is removed and the recess below is either filled with conditioning liquid or left empty. Narrow troughs (3) hold the mobile phase. Development is started by pushing the levers, which tilt the glass strips (4) inward. Solvent travels to the layer through the resulting capillary slit. Development stops automatically when the solvent front meets in the centre (development distance, 4.5 cm). The chamber, which is constructed from poly(tetrafluoroethylene), is kept covered with a glass plate (5) at all times (Permission for use granted by Camag, Muttenz, Switzerland)

Function of the 'U'
Chamber
1. HPTLC plate
2. Base support
3. Solvent delivery inlet
4. Vapour phase inlet
5. Vapour phase outlet

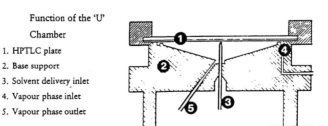

Figure 15 *Cross-section of the U-chamber. Designed for circular chromatograms*

component. This allows for a much more critical measurement of the position of the analytes.

5.4 Automatic Developing Chamber (ADC)

As the name suggests this equipment fully automates the TLC development procedure for the linear ascending technique. Such instruments, like the one shown in Figure 16, provide complete control of the chromatogram development so that all chromatograms are totally reproducible. Layer pre-conditioning, tank or sandwich configuration, solvent migration distance, and drying conditions are all selectable. Parameters can be entered via a keypad and a memory feature allows complete developing programs to be stored for future use. The progress of development is monitored using a CCD (charge coupled device) sensor, and the migration time and distance are displayed continuously. Once the program run is complete the chromatogram is dried in a choice of warm or cool air.

Figure 16 *Camag automatic developing chamber (ADC). All parameters are selectable using the keypad*
(Permission for use granted by Camag, Muttenz, Switzerland)

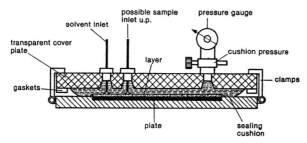

Figure 17 *Cross-section of an over-pressurised (forced flow) TLC system (OPLC)*
(Reprinted from F. Geiss in *Fundamentals of Thin Layer Chromatography*,
1987, p 158, Hüthig, Heidelberg, Germany, with permission of the publisher,
Wiley-VCH and the author)

5.5 Forced-flow Development (OPLC)

This technique began its history in Hungary in the late 1970s with the research of
E. Tyihak *et al.*[17] It is a forced-flow technique, often also called over-pressurised
liquid chromatography (OPLC). The solvent is applied to the sorbent layer by a
constant volume pump permitting the plate length and solvent migration rate for a
separation to be optimised independently of each other. The construction of the
chamber is such that the sorbent layer is sandwiched between a glass backing plate
and a flexible polymer membrane (sealing cushion) forced into close contact with
the layer by application of hydraulic pressure (see Figure 17).

The two basic parts of a forced-flow TLC instrument are the upper and lower
support blocks, which form a sandwich with the thin-layer plate. The bottom
support block may contain a thermostatting jacket. The upper support block holds a
membrane, the mobile phase inlet, pressure inlet, and pressure gauge, as well as an
arrangement for insulating these. The sorbent layer on a glass plate is positioned
facing the membrane. Clamps hold the upper and bottom support blocks together.
The advantages of this forced-flow TLC equipment for separations are listed in
Table 6. Unlike the previous development chambers, the OPLC system requires
mobile phase to be pumped under pressure through the sorbent layer. Hence

Table 6 *Advantages of forced-flow thin-layer chromatography*

1.	The solvent velocity can be optimised independently of other experimental variables
2.	Higher efficiencies can be achieved using fine particle sorbents and longer plates than is possible with capillary controlled flow systems
3.	A linear increase in efficiency with increasing migration distance is obtained. The upper limit is set by the length of the sorbent layer and the pressure required to maintain the optimum linear mobile phase velocity for that length
4.	Analysis time can be reduced by a factor of 5 to 20
5.	Mobile phases with poor sorbent wetting characteristics may be used
6.	Solvent gradients, either step gradients or continuous gradients, can be employed to optimise the separation conditions

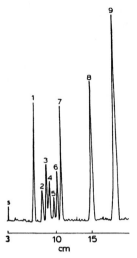

Figure 18 *Spectrodensitometric trace of a standard mixture of doping agents separated by OPLC at external membrane pressure of 1.0* MPa
Sorbent layer: HPTLC silica gel 60 F$_{254}$
Mobile phase: butan-1-ol–chloroform–butanone–water–acetic acid (25:17:8: 4:6 v/v)
Flow rate: 0.85 cm min^{-1}
Detection: UV, 210 nm
Peaks: 1, strychnine; 2, ephedrine; 3, methamphetamine; 4, phenmetrazine; 5, methylphenidate; 6, amphetamine; 7, desopimon; 8, coramin; 9, caffeine
(Reprinted from H Gulyás *et al., J. Chromatogr.,* 1984, **291**, 471–475 with permission of the publisher, Elsevier)

although not shown as a part of the equipment in Figure 17, a suitable solvent pump delivering a non-pulsed flow of the type used in HPLC results in very short development times.

The main advantages of the OPLC technique are reduced development time, although this must be weighed against plate preparation time required, and the increased spot capacity possible. For normal capillary flow HPTLC, spot capacities of 20 to 25 are about the limit. However, present commercial OPLC equipment can achieve spot capacities up to 80 (comparable to column chromatography). Further improvements in the separation number as mentioned previously are only possible by using longer plates, two-dimensional techniques or by automated multiple development (AMD) the latter two being described later. Figure 18 illustrates a typical separation with the OPLC technique.

5.6 Vario Chambers

The Vario chamber has remained an important piece of equipment since the late 1960s for the optimisation of development conditions on TLC and HPTLC plates. There is no doubt that a Vario chamber is a most useful tool for scouting the optimum solvent/solvent mixture, and solvent vapour conditions for a desired separation. Both S-chamber and N-chamber configurations can be simulated using solvent vapour conditioning or non-conditioning formats. As well as allowing the development of several solvents side by side, up to six different conditions of pre-loading, including relative humidity can be tested simultaneously. A single unit to achieve this is therefore very desirable One type of Vario chamber is shown in Figure 19.

As six chromatograms can be run on one plate, the results of attempts to optimise the mobile phase can immediately be compared. Volumes of solvents used are very small as each well in the chamber is designed to accommodate about one millilitre. There is therefore little solvent wastage. The six chromatographic developments usually take about the same time to complete. Hence the time of optimisation of the solvent conditions is dramatically reduced. An example of how the chamber has been used is demonstrated in Figure 20 by optimising a solvent mixture for the separation of corticosteroids.

The Vario chamber from Camag is used in the following way:

1. A TLC/HPTLC glass plate is pre-scored to give six channels. The sample solutions are then applied as spots or bands in the usual way on each sorbent track.
2. If pre-conditioning of the sorbent layer is required, then at this point the conditioning liquid is poured into the troughs of the conditioning block.
3. The chromatographic layer is placed face down on the block with the channels over the troughs and with the sample origins facing the solvent wells. Retaining clips are used to clamp the plate in position.
4. A pipette is used to fill the solvent wells with about one millilitre of developing solvent. Transfer glasses are placed in the wells and tilted outwards.
5. After the desired pre-equilibration time has elapsed, the clips are released, the plate pushed towards the solvent wells, and the plate re-clipped in position. The transfer glasses are tilted against the layer edge and the solvent is transferred on to the TLC layer by capillary attraction. The development then proceeds in the normal way.

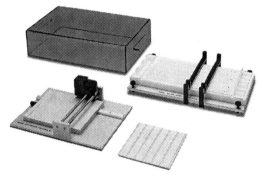

Figure 19 *The Camag HPTLC Vario chamber. The unit is designed for the optimisation of developing parameters in HPTLC. Six developing solvents or pre-equilibrium conditions can be investigated side by side. Developing conditions of both tank and sandwich configurations can be simulated. Combinations of the above can also be tested. The six channels on the HPTLC layer are achieved using the scoring device shown*
(Permission for use granted by Camag, Muttenz, Switzerland)

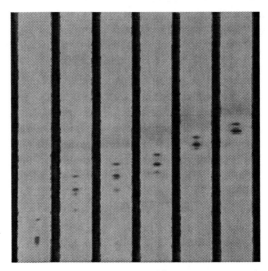

Figure 20 *Optimisation of a developing solvent for the separation of corticosteroids after uniform pre-conditioning at 72% relative humidity. The increase in methanol content increases the polarity of the mobile phase resulting in increased migration rate. Subsequent improvement in resolution of closely related compounds occurs and it can be observed that the resolution power goes through a maximum as the methanol content increases. The optimum separation is achieved with chloroform-methanol (95:5% v/v).*
Sorbent layer: HPTLC silica gel 60
Mobile phase: chloroform-methanol
Ratio: 93:7, 95:5, 90:10, 85:15, and 80:20% v/v
Detection: UV quenching at 254 nm

6 Multiple Development Techniques

Multiple development techniques are in some cases quite unique to TLC. They are also very powerful techniques, focusing separated zones, often improving resolution and spot/band capacity many-fold. Multiple development can be split into several types that will now be described in more detail.

6.1 Single Mobile Phase Multiple and Continuous Development

Single mobile phase multiple development proceeds in the following way. The first development is allowed to proceed until the mobile phase has migrated to a predetermined point on the sorbent layer. The plate is then removed, dried and re-introduced into the same solvent mixture. Development is again allowed to proceed until the mobile phase has migrated the same distance. The process can be repeated until the sample components have migrated sufficiently to be resolved.

This technique is often used in amino-acid analysis where the sample contains many components and a single development is insufficient to cause migration above the lower part of the sorbent layer, and also to separate fully the amino-acids of

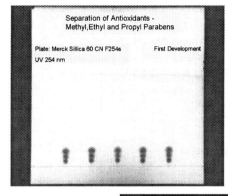
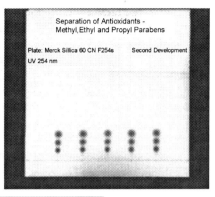
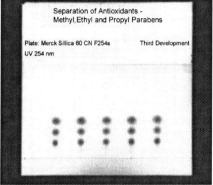

Figure 21 *Improved resolution of analytes by multiple development using the same mobile phase*
Sorbent layer: HPTLC silica gel CN F_{254s}
Mobile phase: ethanol–water (20:80 v/v)
Detection: UV fluorescence quenching at 254 nm
Spots: propyl paraben, ethyl paraben, and methyl paraben (in ascending order)

interest. However, it is by no means limited to just this one application. There are many instances where several developments with the same solvent give improved resolution. One example is shown in Figure 21 where several developments of antioxidants commonly used in beverages shows a dramatic improvement in resolution.

Sometimes a TLC plate does not seem long enough to give the required separation of components of the sample. In these cases the continuous development method can be useful. It also offers a number of other advantages over multiple development. Only one developing solvent system is used. This is allowed to ascend the sorbent layer in the usual way, but provision is made at the top of the layer for evaporation with a "cut off" in the lid of the chromatography chamber. As the solvent evaporates from the top of the sorbent layer, a continuous flow of mobile phase is maintained. Unlike multiple development, reconditioning of the layer, equilibration of the TLC tank, and drying steps before every development are eliminated. Also repetitive multiple development with the same solvent is theoretically and sometimes practically less effective as regards resolution. For the second and all subsequent runs in multiple development, the rising front reaches the lowest spot/band first. As a consequence, this starts to migrate before the solvent front reaches the next spot/band, thus reducing the ΔR_f and detracting from the resolution achieved on the previous run. However, once the front passes the second spot/band the effect readjusts somewhat as this now migrates at a faster rate. For sample components of low R_f value, the solvent front is travelling at its fastest rate and this adverse effect is not important. However, for sample components that have migrated further up the TLC plate, this resolution lessening effect becomes more significant. The number of runs therefore goes through an optimum with respect to separation (as shown in Figure 22). The following equation enables a calculation to be made of the number of developments that can be performed without losing the resolution gained:

$$n_o = \frac{-1}{\log(1 - {}^1R_f)} \tag{11}$$

At low R_f this can be simplified to:

$$n_o = \frac{1}{{}^1R_f} \tag{12}$$

where ${}^1R_f$ is the R_f value after the first development
      n_o is the optimum number of developments

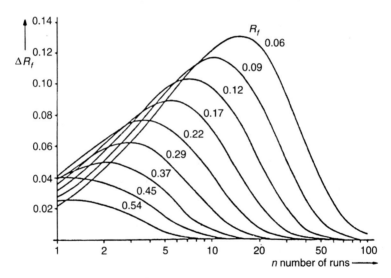

Figure 22   *Multiple development and separation*
         Distance of chromatographic zone centres, ΔR_f versus the number of development runs for single development R_f values. As the R_f value for a single development decreases, the number of runs required to give the maximum ΔR_f increases. In all cases the ΔR_f goes through a maximum value with increase of the development runs
         (Reprinted from F. Geiss in *Fundamentals of Thin Layer Chromatography*, 1987, p 145, Hüthig, Heidelberg, Germany with permission of the publisher, Wiley-VCH and the author)

Figure 22 shows clearly the effect that occurs on multiple development. In practice it is found that the number of runs must be 10–20% higher than the calculated n_o to achieve maximum resolution. There is a relationship between ${}^1R_f$ and ${}^nR_f$ after n runs:

$$^nR_f = 1 - (1 - {}^1R_f)^n \tag{13}$$

However, what is more important is being able to calculate the number of runs (n) necessary to raise ${}^1R_f$ to ${}^nR_f$:

$$n = \frac{\log(1 - {}^nR_f)}{\log(1 - {}^1R_f)} \tag{14}$$

Although continuous development has some advantages over single mobile phase multiple development, the major disadvantage is reproducibility of the chromatogram. As one cannot control the solvent front migration distance or the speed of migration, it is possible to attain a degree of reproducibility based on the time of development.

6.2 Two-dimensional Development

Planar chromatography is the only technique of chromatography where two-dimensional (2-D) development is possible. It is a very powerful separation tool and yet is quite often overlooked as the procedure of choice. Unfortunately most two-dimensional separations in the past have involved the separation of at least 20 amino-acids on cellulose or silica gel, where the procedure takes all day to run and only one sample per plate can be analysed at a time. As the resulting chromatogram is a "fingerprint", identifying the spots by comparison with standards is time consuming as these will need to be run separately under the same mobile phase conditions. However, once a method has been devised, the technique is a very reliable and rugged one. Traditionally, amino-acids have been separated this way for decades, (for a typical separation, see Figure 23).

In order to obtain good resolution, it is necessary to choose two solvent mixtures that are quite different, although of similar solvent strength. This is quite difficult, but important as otherwise the resolution of components will be no better than a linear ascending chromatogram. For the sorbents, cellulose and silica gel, it is best to choose the solvent mixture for the first dimension so that at least half of the suspected components in the sample are separated on a linear chromatogram. The solvent mixture for the second dimension will often need to give similar resolution to that of the first, but in this instance the linear chromatogram will need to be closely compared with the first one before deciding whether it is suitable. The first chromatogram will have shown better resolution of some components than others, whereas the second chromatogram will need to show better resolution of those components poorly separated by the first solvent mixture.

Normally amino-acid separations are achieved by a combination of single solvent multiple development and two-dimensional development. The sample is applied as a spot to the sorbent layer in the corner of the plate, usually 1 cm from each edge. After appropriate drying, the chromatogram is developed in the first dimension with a solvent mixture that is polar, usually basic in nature. The plate is

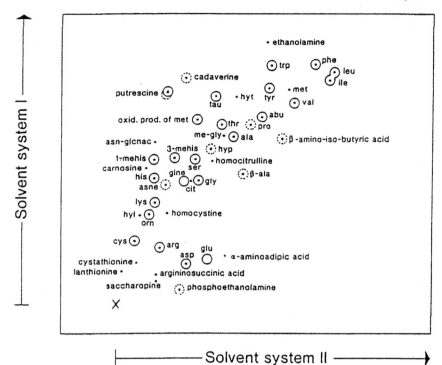

↑ Solvent system I ──────

|────────── Solvent system II ──────→|

Amino Acids
Reference
Solution:

1 μL of solution 1+1) contains:		Solvent: 2-propanol – water,	
α-Alanine	12 μg	Lysine	25 μg
β-Alanine	25 μg	Methionine	25 μg
α-Aminobutyric acid	12 μg	1-Methylhistidine·H₂O	50 μg
β-Aminobutyric acid	40 μg	3-Methylhistidine	50 μg
Arginine·HCl	50 μg	Ornithine.HCl	25 μg
Asparagine	60 μg	Phenylalanine	70 μg
Aspartic acid	25 μg	Phosphoethanolamine	40 μg
Cadaverine	90 μL	Proline	25 μg
Cystine·HCl	85 μg	Putrescine	150 μL
Glycine	12 μg	Serine	12 μg
Histidine·HCl	60 μg	Taurine	60 μg
Homocitrulline	25 μg	Threonine	25 μg
Hydroxyproline	50 μg	Tryptophan	50 μg
Leucine	12 μg	Tyrosine	25 μg
Isoleucine	12 μg	Valine	12 μg

Figure 23 *Two-dimensional separation of amino-acids in urine*
Sorbent layer: cellulose TLC plate, 10×10 cm
Mobile phase 1: pyridine/1,4-dioxan/ammonia soln. (25%)/water (35:35:15: 15 v/v)
Mobile phase 2: ethanol/acetone/acetic acid/water (35:35:7:23 v/v)
Detection: Ninhydrin solution and heated for 15 min. at 120 °C
(By permission of Merck)

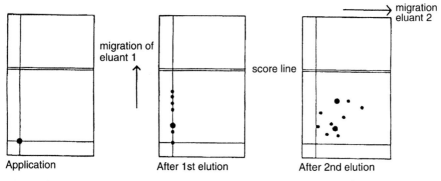

Figure 24 *The procedure of two-dimensional thin-layer chromatography*

then dried and the chromatogram is developed again in the same direction with the same solvent mixture. The plate is then thoroughly dried to remove all solvent vapours, turned through 90° and developed with a different solvent, again a polar mixture, but usually acidic. Again two developments with intermediate drying are necessary to separate large numbers of amino-acids. (Figure 24 shows diagrammatically the technique of two-dimensional separation.)

Typical solvent mixtures that are used for amino-acid separations are shown in Table 7. Amino-acids are visualised with ninhydrin reagent (normally as a spray solution in butan-1-ol).

As the choice of solvents often presents a problem, chromatographers have sought to remove the difficulty by using other stationary phases where changes in

Table 7 *Developing solvent mixtures that have been recommended for two-dimensional TLC separation of underivatised amino-acids*

Sorbent	Solvent Mixture for 1st Dimension	Solvent Mixture for 2nd Dimension
Cellulose	Pyridine/1,4-dioxan/ammonia soln. 25% w/w/water (35:35:15:15 v/v)	Butan-1-ol/acetone/acetic acid/water (35:35:7:23 v/v)
Cellulose	Pyridine/1,4-dioxan/ammonia soln. 25% w/w/water (35:35:15:15 v/v)	Ethanol/acetone/acetic acid/water (35:35:7:23 v/v)
Cellulose	tert-Butanol/butanone/ammonia soln. 25% w/w/water (45:27:9:18 v/v)	Butan-2-ol/acetone/acetic acid/water (21:21:6:12 v/v)
Cellulose	tert-Butanol/butanone/ammonia soln. 25% w/w/water (45:27:9:18 v/v)	Ethanol/acetone/acetic acid/water (21:21:6:12 v/v)
Cellulose	tert-Butanol/butanone/ammonia soln. 25% w/w/water (50:30:10:12 v/v)	Butan-1-ol/acetone/acetic acid/water (35:35:20:10 v/v)
Silica gel 60	Butan-1-ol/propionic acid/water (40:10:10 v/v)	Pentan-1-ol/butanone/pyridine/water (20:20:20:15 v/v)
Silica gel 60	Butan-1-ol/acetic acid/water (40:10:10 v/v)	Pentan-1-ol/butanone/pyridine/water (20:20:20:10 v/v)

mobile phase would make dramatic differences in separation. Two approaches have been used, both based on the same idea:

Dual Phase Layer. A dual phase as described in Chapter 2 allows development in the reversed-phase zone with very polar solvents (*e.g.* methanol/acetonitrile with water), and further separation in the normal-phase part of the plate with much less polar solvents. As one switches from one stationary phase to the other, the order of migration of separated components is reversed.

Cyano-bonded Silica Gel. As described in Chapter 2 this bonded layer has the ability to be used for dual phase partition chromatography. With non-polar solvents it separates on the basis of a normal-phase partitioning mechanism, whereas with polar solvents the mechanism reverts to reversed-phase partitioning. The plate can therefore be used in exactly the same way as cellulose or silica gel 60 layers, but the resolution is noticeably improved.

6.3 Manual Gradient Development

Gradient development is employed where the sample contains both non-polar and polar analytes. In its simplest form a strong solvent is chosen and allowed to migrate halfway up the plate. This solvent is polar enough to cause some migration of the most polar sample components and for the non-polar components to travel with the solvent front. After drying, development is repeated with a weaker solvent, but this time the solvent front is allowed to migrate the entire separation length. The substances separated on the first run remain almost immobile whilst those on the solvent front (less polar ones) now separate on the upper part of the plate. In its usual form, gradient development involves a stepwise development in which the most polar solvent is used to cause migration of all the sample components, but the solvent front is only allowed to travel a short distance. After drying, the process is repeated with a slightly less polar solvent, allowing the solvent front to travel slightly further. This process is repeated 10 or 20 times reducing the polarity of the solvent each time until a satisfactory separation is attained. The system can be further advanced into Automated Multiple Development (AMD) where the whole process of gradient TLC/HPTLC is automated.

6.4 Automated Multiple Development (AMD)

As the name suggests, AMD is based on the same principle as manual multiple development. An AMD chromatogram is the result of several (usually at least 10) automated chromatographic runs. In most cases the number varies between 10 and 40. The migration distance of the solvent front in each run is progressively longer by a constant increment. Typical increments are from 1 to 3 mm. The distance is programmed via computer control of the instrument. As each solvent run is completed, the developing tank is drained and the HPTLC plate dried by a vacuum system. Before the next run begins, solvent vapour is pumped into the chamber to pre-condition the plate. As the early solvent mixtures will be most polar (normal-phase development), the sample components of interest will be focused or sharpened into tight bands (see Figure 25).

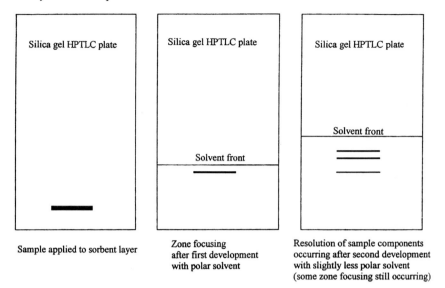

Sample applied to sorbent layer

Zone focusing
after first development
with polar solvent

Resolution of sample components
occurring after second development
with slightly less polar solvent
(some zone focusing still occurring)

Figure 25 *The effect of zone focusing during automated multiple development.*
Here the use of methanol or acetonitrile as the first polar solvent for a short
development distance on silica gel 60 followed by slightly less polar solvents
results in sharpening or focusing the chromatographic zones. This improves the
detection limit and resolution

This occurs because the solvent reaches the lower edge of the chromatographic
zone first. This begins to migrate whilst the upper part of the zone remains in the
dry part of the sorbent. A concentration of the zone into a much thinner band
therefore occurs at or near the solvent front. Once achieved, the later solvent
mixture runs cause little noticeable diffusion into the sorbent layer. Hence
separation numbers of 25 to 30 are routinely observed. An AMD elution gradient
for normal-phase separation therefore starts with a strong, polar solvent and ends
with a weak, non-polar one. As expected for reversed-phase separations, the
opposite procedure for choice of solvent polarity is applied.

Solvent mixtures are prepared by a pump and gradient mixer for pure solvents
contained in storage jars. An example of a commercially available AMD unit is
shown in Figure 26. This system has six solvent jars and is fully programmable,
with the ability to store and recall gradient methods. A typical AMD chromatographic separation of pesticides using this system is shown in Figure 27. It consists
of 10 runs of dichloromethane – acetonitrile (70: 30 v/v) to extract the "active"
components to a migration distance of 10 mm on the plate. This was followed by
5 runs involving step changes to pure dichloromethane. Finally during a further
10 runs a constant change is made to pure n-hexane.

If required, the mobile phase can be adjusted at any time with acid or base
modifiers. Usually methanol or acetonitrile are chosen as the polar solvents,
dichloromethane and tert-butylmethyl ether as the intermediate solvents, and
n-hexane as the non-polar solvent. However, when necessary the polarity can be

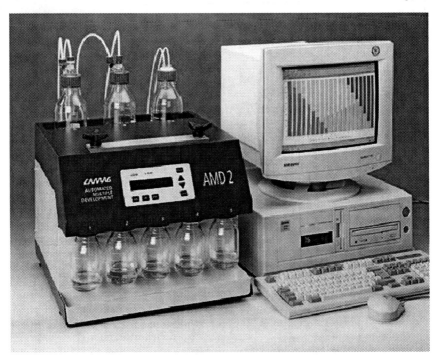

Figure 26 *The Automated Developing System (AMD) manufactured by Camag. The developing chamber is situated in the front portion of the unit. The five reservoir bottles under the front housing hold the solvents for use in the developing cycles. Located at the back of the hood is the gas wash bottle, a molecular sieve vessel for making up the gas phase, and a solvent wash bottle. The computer controls any required pre-conditioning, development steps, migration times, solvent concentrations and drying times*
(Permission for use granted by Camag, Muttenz, Switzerland)

adjusted at any stage to "fine tune" the resolution of the separation. Figure 28 shows another example of a typical AMD separation that demonstrates the improved resolution achievable with this technique.

However, it should be borne in mind that this form of TLC development is not recommended for analytes that are volatile or unstable when subjected to repeated drying. Solvents with high boiling points, low volatility, and high viscosity should be avoided as development and drying times will become very long. In order to improve the migration rates for solvents in AMD, thin coated HPTLC plates have been commercially developed that have a silica gel thickness of 100 or 50 μm. Using these layers, most AMD separations can be completed in 30 to 60 minutes.

Since its beginning the AMD approach to complex separations has proved to be an excellent tool for noticeable improvement in resolution. Examples abound in the TLC literature with separations of carbohydrates (14 components),[18] pesticides (16 components),[19–22] phenolic compounds (13 components),[23,24] vanilla extract (10 components),[25,26] phospholipids (9 components),[27] camomile extract (15 components),[28] gangliosides (9 components)[29] and amines (9 components).[30]

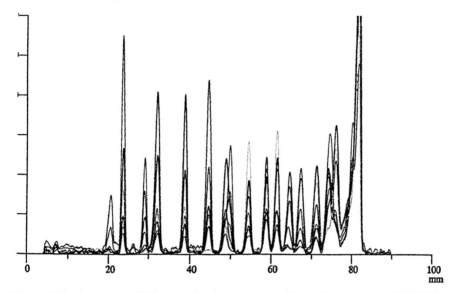

Figure 27   *Separation of 16 pesticides demonstrating the resolving power of AMD. The result is a multi-wavelength scan of one chromatographic track.*
Sorbent layer: HPTLC silica gel 60 F_{254s}, 20×10 cm
Mobile phase: 20-step gradient based on n-hexane/acetonitrile/dichloromethane
Peaks: 2,4-D, atrazine, benzanilide, chlortoluron, cyanazine, desethylatrazine, metazachlor, metobromuron, metoxuron, monuron, phenmedipham, phendimethalin, propazin, simazine, terbutylazine, vinclozolin (25 ng ml^{-1} of each)
(Permission for use granted by Camag, Muttenz, Switzerland)

7 Mobile Phase Additives

7.1 Use of Buffers and pH

Several equations have been derived to show the dependence of pH in TLC on the R_f and R_m values, the phase ratio (A_s/A_m), and the distribution coefficient (K). The important relationship is:

$$R_m = \log K\left(\frac{A_s}{A_m}\right) + pH - pKa \qquad (15)$$

where $R_m = \log k' = \log \dfrac{(1 - R_f)}{R_f}$

Although R_f values are directly related to the separations obtained, the R_m value often proves more convenient in the theory of optimisation of chromatographic systems as it can be directly related to a number of parameters, such as temperature, molecular structure of the analyte, solvent composition, and pH. Of course, the use of pH and buffers can only affect the final chromatogram if the analytes are capable of dissociation in water. One of the domains of pH-controlled separations is in ion-exchange TLC. Unfortunately few strong anion and cation pre-coated

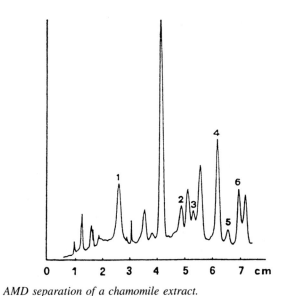

Figure 28   *AMD separation of a chamomile extract.*
Sorbent layer: silica gel 60 HPTLC, 10×20 cm glass plate
Mobile phase: 15-step gradient based on methanol/dichloromethane/water/
formic acid (from 70.5:25:4.5:1 to 0:100:0:0 v/v)
Detection: fluorescence with excitation at 360 nm
Peaks: 1, apigenin 7-O-glucoside; 2, caffeic acid; 3, apigenin; 4, umbellifer-
one; 5, ferulic acid; 6, herniarin
(Reprinted from E. Menziani *et al., J. Chromatogr.*, 1990, **511**, 396–401 with
permission of the publisher, Elsevier)

plates are commercially available to utilise this technique. However, weak base
anion-exchangers do exist in the form of amino-bonded and PEI cellulose TLC
layers. These were described in Chapter 2 along with examples that clearly
demonstrate the types of separation that are achievable.

Figure 29a   *Complex formation of carbohydrates with boric acid. Where acid conditions*
prevail in the mobile phase (low pH), complex I is formed. However, if the
conditions are alkaline (high pH), further reaction occurs to form complex II
Figure 29b   *Typical separation of carbohydrates in honey using 2-aminoethyldiphenyl*
borate as complexing reagent to improve resolution.
Sorbent layer: silica gel 60 HPTLC, 10×10 cm glass plate
Mobile phase: acetonitrile/15 mM sodium dihydrogen phosphate buffer, pH
5.5 (85:15 v/v)
Complexing agent: 0.034 g/100 ml 2-aminoethyldiphenyl borate
Chamber: fully saturated N-chamber
Visualisation: 2% w/v diphenylamine and 2% w/v aniline in methanol –
orthophosphoric acid (80:20 v/v).
Tracks: 1, maltose; 2, sucrose; 3, glucose; 4, fructose; 5, honey; 6, golden
syrup (overloaded); 7, golden syrup
Concentrations: all standards, 1000 ppm, honey, 2000 ppm, golden syrup,
2000 ppm and 200 ppm respectively

A

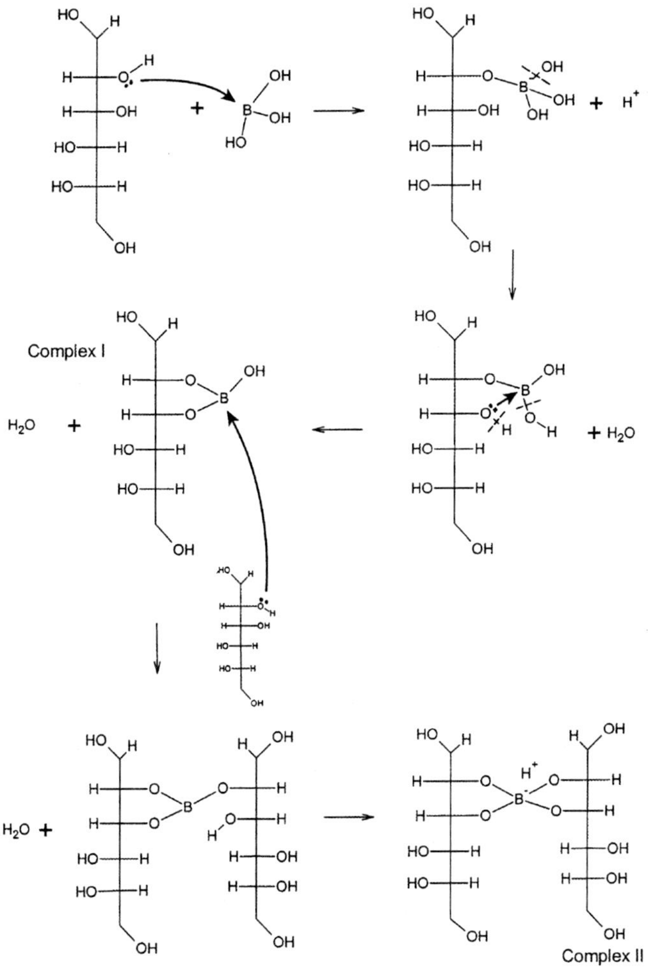

B

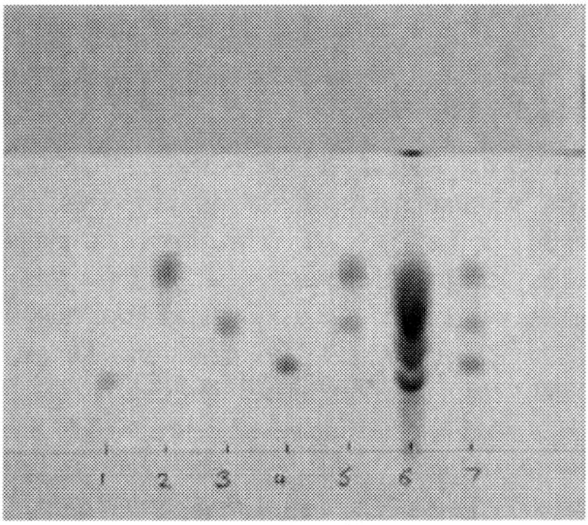

One specific area of separation where buffers have proved invaluable in obtaining improved resolution of analytes is in carbohydrate analysis. In Chapter 2 impregnation of silica gel layers with sodium or potassium dihydrogen phosphate was described. Addition of these buffers to the developing solvent often results in the same or similar effect. Typically 50 mM concentrations will be sufficient to inhibit glycamine formation from reducing sugars (the principal separation application for this technique). The use of phosphate and borate buffers either in the developing solvent mixture or by impregnation into the sorbent layer results in clearly resolved carbohydrates. The borate radical reacts readily with polyols present in the carbohydrate structure to form unstable ionic complexes (see Figure 29a). This effect appears to be most pronounced where silica gel is the sorbent. An example of the results that can be obtained is shown in Figure 29b. This particular separation incorporates the use of 2-aminoethyldiphenyl borate in the developing solvent mixture as the complexing agent. Well resolved analytes and good reproducible results are achieved for carbohydrate standards, honey, golden syrups and similar products.

7.2 Acid or Base Additives

The use of acid and base additives is really an extension of the pH effects described earlier. In many cases separations can be improved by the addition of an acid, usually acetic, formic or propionic, or a base, such as ammonia solution, triethylamine or pyridine. These may have the effect of reducing "tailing", resulting in sharper, less diffuse, and more circular spots or even dramatically altering the retention rate and sometimes changing the order of migration of the components of the sample. Most of these results can be accounted for by the nature of the analytes. If the molecules are charged or by altering the pH, they can be ionised or the ionisation suppressed, then addition of acid or base most often causes the above changes and noticeably improves the chromatographic separation. This illustrates the need to obtain as much information about the nature of the sample before chromatography. Examples of the use of acid or base additives are shown in Figures 7,8 in Chapter 2.

7.3 Ion-pairing Reagents

The use of ion-pairing reagents has proved successful in both improving the quality of separations and in obtaining resolution of compounds that hitherto had been difficult. The method relies upon the formation of a complex between ions of opposite charge, the solute and the counter-ion. For ion-pairing to occur, the sample components must be ionised and a suitable ion-pairing reagent chosen of opposite charge. pH therefore again plays an important part. For example, a sample may contain components that are completely ionised at pH 7 where ion-pairing formation will be at a maximum. However, when the pH is lowered to 4, sample anions begin to form un-ionised acids, R_f values decrease, and due to the variation in ionisation between components, ion-pairing formation will vary and separation will occur. As a general rule in reversed-phase systems if the pKa of an analyte is 5, then it will migrate faster as the pH drops below about 5.5. For an analyte with a pKa of 3, little change in the migration rate will occur until the pH drops below about 3.5. Conversely for normal-phase systems, the results are inverted.

$$\text{Analyte}^{(-)} + \text{Counter-ion}^{(+)} = \text{Ion-pair}^{(\pm)}$$

$$\text{Analyte}^{(+)} + \text{Counter-ion}^{(-)} = \text{Ion-pair}^{(\pm)}$$

e.g.

$$R' - SO_3^{(-)} + (C_4H_9)_4N^{(+)} = [(C_4H_9)_4N] - (R' - SO_3)$$

$$R_4N^{(+)} + C_5H_{11}SO_3^{(-)} = (R_4N) - (C_5H_{11}SO_3)$$

In TLC/HPTLC the reagent can be incorporated into the separation sequence in two ways:

1. The sorbent layer is impregnated with a solution of the ion-pairing reagent (usually prepared in as volatile a solvent mixture as possible without precipitation, at a concentration of 0.001–0.1 M). After drying, the plate is ready for the application of the sample and development in the desired solvent.

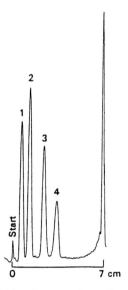

Figure 30 *Separation of alkaloids using an anion-pair reagent in the mobile phase.*
Sorbent layer: silica gel 60 RP18 F$_{254s}$ TLC, 10×20 cm glass plate
Mobile phase: methanol/50 mM potassium dihydrogen phosphate, pH 4 (40:60 v/v)
Ion pair: 100 mM octane sulphonic acid sodium salt
Chamber: N–chamber, without saturation
Peaks: 1, eupaverine chloride (0.2%); 2, papaverine (0.3%); 3, caffeine (0.3%); 4, codeine (1.0%)
Detection: scanning in absorption mode at 254 nm
(By permission of Merck)

2. The sample is applied to the sorbent layer in the usual way and the ion-pairing reagent incorporated into the mobile phase, normally at an overall concentration of 1–2% w/v.

Research has shown that the former approach often gives the most predictable results. For some long chain ion-pair reagents (*e.g.* cetrimide), prior impregnation of the plates with reagent is essential for effective ion-pairing.[31] Even with shorter chain reagents, (*e.g.* tetramethylammonium bromide) there is a noticeable higher degree of ion-pair formation with pre-chromatographic impregnation compared with mobile phase addition. Anion-pairing reagents have been used successfully for the separation of alkaloids,[32] phenothiazine, phenothiazine sulphoxide,[33] and basic dyes.[34] Cation-pairing reagents have similarly been used for alkaloids and for aromatic acids.[35] (For examples, see Figures 30–32.)

7.4 Chiral Additives

In recent years a number of additives to the mobile phase have proved successful for the TLC separation of a substantial number of racemates. Of these the most thoroughly investigated have been ligand-exchange (see also Chapter 2), "Pirkle" reagents, cyclodextrins, cellulose triacetate, BSA (bovine serum albumin), and other

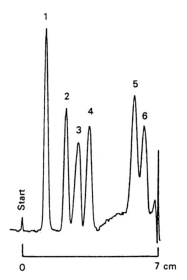

Figure 31 *Separation of nitrogen based compounds using a cation-pair reagent.*
Sorbent layer: silica gel 60 RP18 F₂₅₄ₛ TLC, 10×20 cm glass plate
Mobile phase: acetone/water (50:50 v/v)
Ion pair: 0.1 M tetraethylammonium bromide
Chamber: N–chamber, without saturation
Peaks: 1, methaqualone (0.5%); 2, phenobarbital (0.1%); 3, diphenhydramine (0.3%); 4, barbital (1.0%); 5, codeine (0.2%); 6, morphine (0.1%)
Detection: scanning in absorption mode at 254 nm
(By permission of Merck)

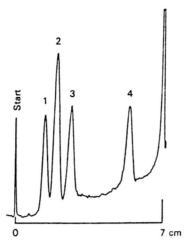

Figure 32 *Separation of organic acids using a cation-pairing reagent.*
Sorbent layer: silica gel 60 RP8 F_{254s} TLC, 10×20 cm glass plate
Mobile phase: acetone/water (40:60 v/v)
Ion pair: 0.005 M tetrapropylammonium bromide
Chamber: N–chamber, without saturation
Peaks: 1, 2-methylbenzoic acid (0.5%); 2, benzoic acid (1.0%); 3, acetyl-salicylic acid (0.5%); 4, phthalic acid (1.0%)
Detection: scanning in absorption mode at 254 nm
(By permission of Merck)

specific chiral selectors. Our interest here in mobile phase choice will limit us to the discussion of these chiral additives where they are introduced into the mobile phase. At present this applies to cyclodextrins, BSA, and other specific chiral additives.

7.4.1 Cyclodextrins

Although the use of cyclodextrin phases in TLC was reported in 1986,[36] it has become more popular to use reversed-phase TLC plates and add the cyclodextrin (usually β-) to the mobile phase.[37] Satisfactory resolution of racemic analytes, such as dansyl DL-glutamic acid can be obtained with concentrations greater than 0.04 M β-cyclodextrin, but for optimum resolution the concentration should be between 0.08 and 0.12 M. High concentrations cause the spots or bands to merge near the solvent front and hence the resolution deteriorates.

Unfortunately the use of cyclodextrins does have a few major drawbacks in TLC. Development times are long, usually 6–8 hours, and cyclodextrins are only slightly soluble in water (1.67×10^{-2} M at 25 °C). Solubility can be increased by dissolving the cyclodextrin in saturated urea solution; concentrations up to 0.2 M are then possible. Modified cylclodextrins, hydroxypropyl-, hydroxyethyl-, and maltosyl-β-cyclodextrin are much more soluble in water (<0.4 M). It has been noted that for hydroxypropyl- and hydroxyethyl-β-cyclodextrins, the solubility increases with the degree of substitution. However, higher concentrations are necessary to achieve the same resolution. This unfortunately results in high

Table 8 *pH dependency of the developing solvent on the retention and resolution of enantimeric tryptophans on Sil C_{18}-50 UV_{254} plates*
(Reproduced by permission of the publishers of the Journal of Planar Chromatography)

DL-*Tryptophans*	Eluent pH 9.30				Eluent pH 9.55			
	R_{l1}[a]	R_{l2}[b]	α[c]	R_s[d]	R_{l1}	R_{l2}	α	R_s
Tryptophan	0.58	0.58	1.00	–	0.58	0.58	1.00	–
Tryptophanamide	0.27	0.27	1.00	–	0.27	0.27	1.00	–
α-Methyltryptophan	0.52	0.52	1.00	–	0.52	0.52	1.00	–
1-Methyltryptophan	0.35	0.35	1.00	–	0.35	0.35	1.00	–
4-Methyltryptophan	0.47	0.58	1.54	2.3	0.45	0.58	1.68	2.5
5-Methyltryptophan	0.42	0.58	1.90	3.7	0.40	0.58	2.07	2.7
6-Methyltryptophan	0.58	0.63	1.23	1.0	0.58	0.63	1.23	1.0
7-Methyltryptophan	0.41	0.50	1.43	2.3	0.40	0.49	1.44	2.3
5-Hydroxytryptophan	0.66	0.66	1.00	–	0.67	0.67	1.00	–
5-Methoxytryptophan	0.42	0.49	1.32	1.7	0.40	0.46	1.28	1.6
Glycyltryptophan	0.62	0.67	1.24	1.1	0.61	0.68	1.35	1.2

DL-*Tryptophans*	Eluent pH 9.80				Eluent pH 9.92			
	R_{l1}	R_{l2}	α	R_s	R_{l1}	R_{l2}	α	R_s
Tryptophan	0.59	0.72	1.77	2.4	0.64	0.76	1.80	2.5
Tryptophanamide	0.31	0.40	1.48	1.4	0.35	0.42	1.34	1.3
α-Methyltryptophan	0.52	0.52	1.00	–	0.55	0.55	1.00	–
1-Methyltryptophan	0.36	0.36	1.00	–	0.42	0.42	1.00	–
4-Methyltryptophan	0.42	0.65	2.56	4.5	0.48	0.70	2.52	4.1
5-Methyltryptophan	0.37	0.61	2.53	2.7	0.43	0.66	2.56	3.2
6-Methyltryptophan	0.66	0.78	1.82	2.2	0.69	0.84	2.32	2.3
7-Methyltryptophan	0.38	0.42	1.18	1.2	0.45	0.45	1.00	–
5-Hydroxytryptophan	0.66	0.66	1.00	–	0.74	0.74	1.00	–
5-Methoxytryptophan	0.41	0.43	1.08	0.8	0.47	0.47	1.00	–
Glycyltryptophan	0.62	0.77	2.05	1.7	0.68	0.77	1.57	1.5

[a] more retained Isomer
[b] less retained Isomer
[c] $\alpha = \dfrac{1-R_{l1}}{R_{l1}} \Big/ \dfrac{1-R_{l2}}{R_{l2}}$
[d] $R_s = 2 \times$ (distance between spots)/(sum of widths of spots)
Mobile phase: 50 mM di-sodium tetraborate containing propan-2-ol (6% v/v) and BSA (6% w/v)

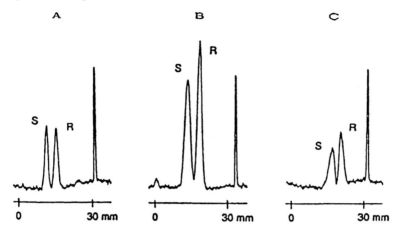

Figure 33 *Separation of the enantiomers of metoprolol (A), propranolol (B), and*
alprenolol (C) using ZGP as additive on moisture equilibrated silica gel 60
DIOL HPTLC, 10×10 cm glass plates.
Mobile phase: dichloromethane stabilised with ethanol
Chiral selector: 5 mM ZGP in the mobile phase
Samples applied: (A) 1 μg, (B) 0.5 μg, (C) 1 μg
Detection: scanning in absorption mode at 254 nm
(Reprinted from A-M Tivert, A Bachman, *J. Planar Chromatogr.*, 1993, 6, 217,
with permission of the editor of Journal of Planar Chromatography)

viscosity mobile phases and long development times (often in excess of 40 hours).
The results with the maltosyl- derivative are similar.[38,39]

7.4.2 Bovine Serum Albumin (BSA)

Racemic tryptophan, substituted tryptophans, and dinitro-pyridyl, phenyl, and
benzoyl amino-acids can be resolved into their respective enantiomers on reversed-
phase TLC plates using high concentrations of Bovine Serum Albumin (BSA) in
the mobile phase. The resolution is very dependent on pH, with the best results
obtained at pH 9–10. The amount of propan-2-ol in the mobile phase also plays an
important part in the separation as does the type of reversed-phase layer. (Some
layers cannot be used at high pH [<9.55].) Optimum eluents contain 6% w/v BSA
and propan-2-ol. The pH dependency is shown in Table 8.[40]

7.4.3 Other Chiral Selectors

N-Carbobenzyloxyglycyl-L-proline (ZGP). The chiral separation of the β-blocking
drugs, alprenolol and propanolol was reported in 1989 using ZGP in the mobile
phase.[41] Since then the procedure has further been used for the resolution of
enantiomers of phenylpropanolamine, pindolol, norphenylephrine, isoproterenol,
and timolol.[42] Both silica gel and diol-modified silica have served as satisfactory
stationary phases with developing solvents consisting of dichloromethane, and
dichloromethane mixed with methanol or propan-2-ol. The concentration of ZGP
varied from 5–7 mM. However, a basic modifier was always added to achieve

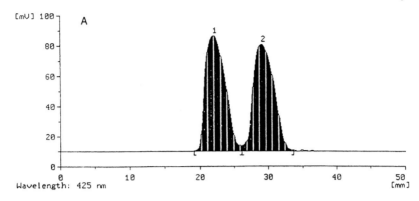

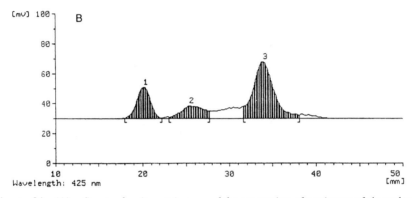

Figure 34   (A) – *Spectrodensitometric scan of the separation of a mixture of* cis *and* trans
isomers of capsaicin.
Sorbent layer: silica gel 60 RP18W HPTLC, 10×10 cm *glass plate*
Sample: 100 nl *containing 0.02%* w/v *capsaicin in dichloromethane*
Mobile phase: methanol/water (50:50 v/v*) + 1.5%* w/v *silver nitrate*
Chamber: N-chamber, fully saturated
Visualisation: 2,6-dichloroquinone-4-chlorimide in ethanol (0.1% w/v*)*
(B) – *Spectrodensitometric scan of the separation of a typical processed chilli*
extract. Chromatographic conditions are as above
Peaks: 1, homodihydrocapsaicin; 2, dihydrocapsaicin; 3, trans-capsaicin
(Reprinted from P. E. Wall, *J. Planar Chromatogr.*, 1997, **10**, 8, with permission
of the editor of Journal of Planar Chromatography)

acceptable resolution; either ethanolamine (0.4 M) or triethylamine (5 mM) has
proved successful. Further research has shown that the resolution can be improved
by increasing the water content of the stationary phase (the diol modified layer) up
to 80% relative humidity (see Figure 33).[43]

(1R)-(-)-Ammonium-10-camphorsulphonate (CAS). This additive has been used in
a similar way to ZGP, with concentrations varying from 6–10 mM in the mobile
phase. The following racemates have been resolved, octopamine, norphenylephrine,
propranolol, isoproterenol, metoprolol, and timolol.[44] However, unlike ZGP
the solubility is poor, limiting the range of solvents which can be selected for

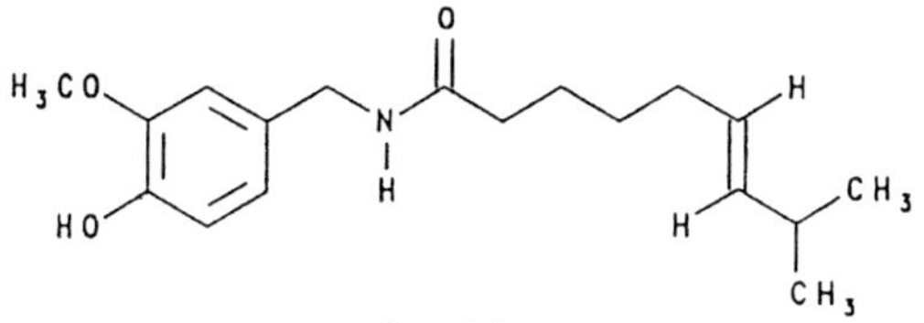

trans-Capsaicin

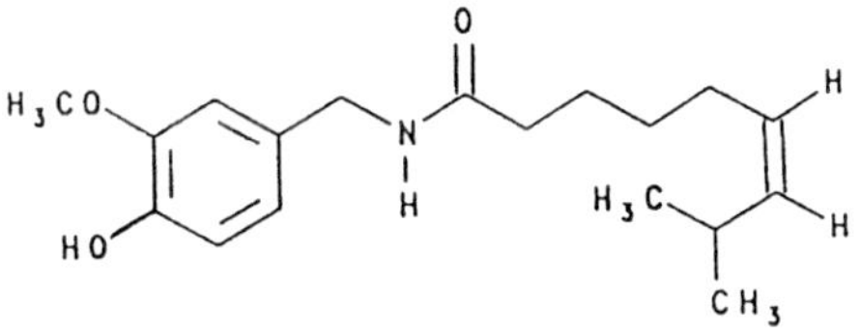

cis-Capsaicin

Figure 35   *The structures of* cis *and* trans *isomers of capsaicin*

Figure 36   *Separation of* cis *and* trans *isomers (triolein and trielaidin) of a C18:1 triacylglyceride using argentation HPTLC.*
Sorbent layer: silica gel 60 RP18 HPTLC, 10 × 10 cm glass plate impregnated with silver nitrate
Mobile phase: dichloromethane/methanol/ethyl acetate/acetic acid/saturated silver nitrate soln. (25:35:20:12:6 v/v)
Chamber: N-chamber, fully saturated
Visualisation: phosphomolybdic acid reagent
Tracks: 1, trans *isomer; 2,* cis *isomer; 3, mixture of* cis *and* trans *isomers*

the mobile phase. CAS is only slightly soluble in dichloromethane, but this can be improved by the addition of at least 10% methanol or propan-2-ol.

7.5 Argentation (Silver Nitrate) TLC

Normally argentation TLC is achieved by incorporating silver nitrate into the sorbent layer by impregnation (Chapter 2). However, the silver nitrate can also be added to the mobile phase in reversed-phase separations where the solvent is aqueous based. This technique has been applied to the separation of *cis* and *trans* isomers of capsaicin and also the homologues of capsaicin on reversed-phase silica gel plates (see Figure 34).[45,46]

Capsaicin and its analogues are the pungent principals of many types of red and green peppers. The resolution is such that quantitative determinations can be carried out on each geometric isomer by *in situ* densitometric scanning of the HPTLC layer. The structures shown in Figure 35 indicate the position of the π-bond in each case. The *trans* isomer exhibits greater retention on the layer than the *cis* isomer due to the position of the $-CH(CH_3)_2$ group. In nature only the *trans* isomer has so far been found to exist.

Similar separations have also been achieved for *cis* and *trans* isomers of unsaturated triacylglycerides as shown in Figure 36. Here a combination of silver nitrate impregnation of the chromatographic layer and modification of the developing solvent have been used to resolve two triacylglycerides that differ only in their *cis/trans* configuration about the π-bonds in the ester chains.

8 References

1. F. Geiss in *Fundamentals of Thin-layer Chromatography (Planar Chromatography)*, Hüthig, Heidelberg, Germany, 1987, 16–20.
2. G. Guiochon and A. Siouffi, *J. Chromatogr., Sci.*, 1978, **16**, 598.
3. A.J.P. Martin and R.L.M. Synge, *J. Biochem.*, 1941, **35**, 1358.
4. A.J.P. Martin and R.L.M. Synge, *J. Biochem.*, 1952, **50**, 679.
5. J.C. Touchstone in *Practice of Thin Layer Chromatography*, 3rd edn., J. Wiley, Chichester, UK, 1992, 94.
6. J. Jacques and J.P. Mathieu, *Bull. Soc. Chem. France*, 1946, **94**.
7. L.R. Snyder, *J. Chromatogr.*, 1964, **16**, 55–88.
8. L.R. Snyder, *J. Chromatogr.*, 1966, **25**, 274–293.
9. L.R. Snyder and J.L. Glajch, *J. Chromatogr.*, 1981, **214**, 1–19.
10. L.R. Snyder and J.L. Glajch, *J. Chromatogr.*, 1981, **214**, 21–34.
11. L.R. Snyder and J.L. Glajch, *J. Chromatogr.*, 1982, **248**, 165–182.
12. L.R. Snyder and J. J. Kirkland in *Introduction to Modern Liquid Chromatography*, 2nd edn., Wiley-Interscience, New York, USA, 1979, 73.
13. L.R. Snyder, *J. Chromatogr.*, 1974, **92**, 223–230.
14. Sz. Nyiredy, C.A.J. Erdalmeier, B. Meier and O. Sticher, *Planta Med.*, 1985, **51**, 241.
15. Sz. Nyiredy, B. Meier, K. Dallenbach-Toelke and O. Sticher, *J. Chromatogr.*, 1986, **365**, 63–71.

16. Sz. Nyiredy, K. Dallenbach-Toelke and O. Sticher, *J. Planar Chromatogr.*, 1988, **1**, 336–342.
17. E. Tyíhák, H. Kalász, E. Mincsovics and J. Nagy, *Proc. Hung. Annu. Meet. Biochem.*, Kreskerńt, 1977, **17**, 183.
18. G. Lodi, A. Betti, E. Menziani, V. Brandolini and B. Tosi, *J. Planar Chromatogr.*, 1994, **7**, 29–33.
19. K. Burger, J. Köhler and H. Jork, *J. Planar Chromatogr.*, 1990, **3**, 504–510.
20. U. de la Vigne, D.E. Jänchen and W.H. Weber, *J. Chromatogr.*, 1991, **553**, 489–496.
21. U. de la Vigne, and D.E. Jänchen, *J. Planar Chromatogr.*, 1990, **3**, 6–9.
22. G.E. Morlock, *J. Chromatogr. A*, 1996, **754**, 423–430.
23. E. Menziani, G. Lodi, A. Bonora, P. Reschiglian and B. Tosi, *J. Chromatogr.*, 1990, **511**, 396–401.
24. G. Lodi, A. Betti, E. Menziani, V. Brandolini and B. Tosi, *J. Planar Chromatogr.*, 1991, **4**, 106–110.
25. M.T. Belay and C.F. Poole, *Chromatographia*, 1993, **37**, 365–373.
26. C.F. Poole and S.K. Poole, *Anal. Chem.*, 1994, **66**, 27A.
27. U. de la Vigne and D.E. Jänchen, *Int. Lab.*, 1991, Nov./Dec., 22–29.
28. W. Markowski and G. Matysik, *J. Chromatogr.*, 1993, **646**, 434–438.
29. Camag Application Note *A-61.1*, Camag, Muttenz, Switzerland.
30. Camag Application Note A-80.1, Camag, Muttenz, Switzerland.
31. S. Lewis and I.D. Wilson, *J. Chromatogr.*, 1984, **312**, 133–140.
32. D. Volkmann, *J. HRC and CC*, 1981, **4**, 350–351.
33. D. Volkmann, *J. HRC and CC*, 1979, **2**, 729–732.
34. P.E. Wall in *Recent Advances in Thin-layer Chromatography,* F. A. A. Dallas, H. Read, R.J. Ruane and I. D. Wilson (eds), Plenum, New York, USA, 1988, 207–210.
35. G.P. Tomkinson, I.D. Wilson and R.J. Ruane, *J. Chromatogr.*, 1990, **3**, 491–494.
36. A. Alak and D.W. Armstrong, *Anal. Chem.*, 1986, **58**, 582–584.
37. D.W. Armstrong, F.-Y. He and S.M. Han, *J. Chromatogr.*, 1988, **448**, 345–354.
38. D.W. Armstrong, J.R. Faulkner and S.M. Han, *J. Chromatogr.*, 1990, **452**, 323–330.
39. J.D. Duncan and D.W. Armstrong, *J. Planar Chromatogr.*, 1990, **3**, 65–67.
40. L. Lepri, V. Coas, P.G. Desideri and A. Zocchi, *J. Planar Chromatogr.*, 1992, **5**, 234–238.
41. A.-M. Tivert and A. Bachman, *J. Planar Chromatogr.*, 1989, **2**, 472–473.
42. J.D. Duncan, *J. Liq. Chromatogr.*, 1990, **13**, 2737–2755.
43. A.-M. Tivert and A. Bachman in *Proceedings of the Sixth International Symposium on Instrumental Planar Chromatography*, H. Traitler, O.I. Voroshilova and R.E. Kaiser (eds), Institute for Chromatography, Bad Dürkheim, Germany, 1991, 19–22.
44. J.D. Duncan, D.W. Armstrong and A.M. Stalcup, *J. Liq. Chromatogr.*, 1990, **13**, 1091–1103.
45. P.E. Wall, *J. Planar Chromatogr.*, 1997, **10**, 4–9.
46. T. Suzuki, T. Kawada and K. Iwai, *J. Chromatogr.*, 1980, **198**, 217–223.

CHAPTER 6

Detection and Visualisation

1 Introduction

Once development of the chromatogram is complete, usually some kind of visualisation of the chromatographic zones is required as most compounds are not visibly coloured. Many compounds will absorb UV light or exhibit fluorescence when excited by UV or visible light, but most require visualisation using an appropriate spraying or dipping reagent. Due to the inert nature of the sorbents commonly used in TLC layers, chemical reactions can be carried out on the plate without the sorbent layer being affected. A wide variety of chemical reagents have been used to good effect for chromatographic zone detection. Even quite aggressive reagents, like hydrochloric or sulfuric acid solutions, can usually be tolerated. The list of visualisation reagents for TLC is large, reflecting the versatility of detection techniques. A number of excellent publications exist in which the formulations for both spraying and dipping reagents have been collated from the chromatography literature.[1-3] Some reagents are termed universal reagents as they are used to visualise a wide range of compounds of differing molecular structures. Included in this group of reagents are acid solutions and vapours, ammonia vapour, fluorescein, dichlorofluorescein, and iodine. Some reagents can be used in destructive techniques. They cause a breakdown of the compounds of interest leaving behind a visible deposit on the chromatographic layer. Conversely there are non-destructive techniques that allow detection of compounds in the chromatographic zones without the layer or zones being chemically altered. Included in non-destructive techniques are visible and UV light, and sometimes the use of iodine or ammonia vapour. The latter two reagents are included as in many cases the "reaction" is reversible. Other reagents are functional group specific and can be used to detect classes of compounds, such as alcohols, aldehydes, ketones, esters, or acids. These are termed group specific reagents. This is about the limit of specificity as no genuine substance specific reagents exist.

Often, separated compounds can be detected and visualised by a combination of the above techniques. A non-destructive technique, such as UV irradiation, may be used first, followed by a universal reagent, and then finally a group specific method to enhance selectivity and sensitivity. Often for a particular analyte there may be several visualisation reagents available. However, there is often a difference is specificity and usually a difference in the limit of sensitivity. Stability also plays an important part in the selection of a suitable detection reagent. Some reagents have good stability over a number of weeks whilst others need to be made up just prior to use. The visualised chromatographic zones may also differ in stability. Some fade quite quickly, whilst others although remaining stable, become more difficult to visualise as the background darkens or is affected in some other way by the reagent. Fortunately the majority of reagents give acceptably stable results. Sometimes dark or coloured backgrounds can be lightened by exposure of the chromatographic layer to acidic or alkaline vapours.

The choice of solvent used to prepare the visualisation reagent also requires consideration. Sometimes a poor selection of solvent results in the visualised chromatographic zones becoming diffuse and developing tails. The effect takes on the appearance of "leaking" zones on the layer. It is usually caused by the analytes in the chromatographic zones being soluble in the reagent solvent. The problem can be solved by using a solvent with a lower elution strength in the preparation of the reagent. However, all these effects will need to be taken into consideration so that the most effective visualisation procedure is used.

2 Non-destructive Techniques

2.1 Visible Detection

Some compounds are sufficiently coloured, for example, natural and synthetic dyes, and nitrophenols, to give an absorption in the visible part of the electromagnetic spectrum. These are clearly seen in visible light and do not require any further treatment for visualisation. The majority, however, as mentioned earlier do require some means of rendering them visible. Hence the need for the following techniques.

2.2 Ultra-violet Detection

Separated chromatographic zones on a TLC/HPTLC layer may appear colourless in normal light, but can absorb electromagnetic radiation at shorter wavelengths. These are often detected in the UV range, normally at 200–400 nm. Often exposure to UV light at short wave radiation (254 nm) or long wave radiation (366 nm) is all that is necessary for absorbing or fluorescing substances to be observed. Most commercial UV lamps and cabinets function at either or both of these wavelengths. Chromatographic zones normally appear dark on a lighter background or if fluorescence occurs, a variety of visible spectrum colours are seen. To aid visualisation many commercial pre-coated TLC layers contain an inorganic phosphorescent indicator, or an organic fluorescent indicator (see Chapter 2). Detection by absorbance in these cases relies on the phosphorescence or fluorescence being quenched by the sample components. This process is commonly called "fluorescence quenching" in both cases, although more accurately for most indicators designated F_{254} it should be described as phosphorescence quenching. In these instances the fluorescence remains for a short period after the source of excitation is removed. (It is very short lived, but longer than 10^{-8} seconds.)[4] Hence, the effect is best observed by continual exposure to UV light.

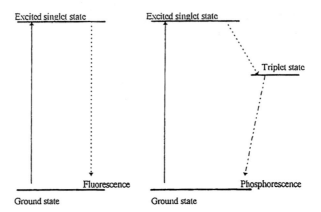

Figure 1   *Electronic transitions during luminescence phenomena*

The process of fluorescence is caused by the electromagnetic radiation providing the energy to bring about an electronic transition from the ground state to an excited singlet state (see Figure 1). As the excited electrons return to the ground state, they emit the energy at a longer wavelength, usually in the visible range. As the diagram shows in Figure 1, phosphorescence is slightly different. Rather than a direct return to the ground state, the electrons return via an excited triplet state to the ground state. As before, the energy is emitted at a longer wavelength. However, as the decay is longer after the source of UV light energy is removed, the process is better described as phosphorescence rather than fluorescence. However, whether the excitation of the indicator in the chromatographic layer exhibits fluorescence or phosphorescence, it is the analyte on the layer that disturbs this process by absorbing the excitation energy. This then results in the typical dark chromatographic zones of the analytes being observed on the fluorescent or phosphorescent background.

Where analytes naturally fluoresce when exposed to UV light, a chromatographic layer containing a fluorescent or phosphorescent indicator may or may not be an advantage. Most of these analytes fluoresce at 366 nm and not at 254 nm. Hence a TLC/HPTLC plate containing an indicator that fluoresces at 254 nm can be an advantage, as both fluorescence and fluorescence quenching can be observed depending on the wavelength. However, if there are analytes present that fluoresce only at 254 nm, or at both wavelengths, then a chromatographic layer without an indicator should be used to avoid background interference. It is also possible to stabilise and sometimes enhance the phosphorescence by the use of certain reagent treatments. A list of the common reagents used is given in Table 1 along with the estimated level of enhancement possible. The effect is simply achieved by dipping the sorbent layer in the reagent solution for a few seconds following treatment with the detection reagent. Normal air drying at room temperature is sufficient to complete the process.

Many analytes, however, do not absorb visible or UV light, quench fluorescence, or fluoresce when excited by visible or UV light. In these instances suitable detection reagents are used to give coloured chromatographic zones in visible light

Table 1 *Some fluorescence intensifiers and their application areas*

Intensifier	Compounds detected	Enhancement	Stabilisation
Triton X-100* (1% v/v solution in hexane or heptane)	Polycyclic aromatic hydrocarbons	10-fold	Partially
Triton X-100 (1% v/v solution in hexane or heptane)	Fatty acids as dansyl amides	At least 10-fold	Yes
Polyethylene glycol 400 or 4000 (10% w/v in methanol)	Compounds with alcoholic (–OH) functional groups	20 to 25-fold	Unknown
Dodecane (50% w/v in hexane)	Polycyclic aromatic hydrocarbons	2-fold	Unknown
Paraffin liquid (33% v/v in hexane)	Aflatoxins	3 to 4-fold	Unknown
Paraffin liquid (33% v/v in hexane)	Ketosteroids, cholesterol, cortisol	10-fold	Unknown
Paraffin liquid (33% v/v in hexane)	Dansyl amides	10-fold	Yes
Paraffin liquid (33% v/v in hexane)	Gentamycins	Yes, but level unknown	Yes

*Triton is a registered trade name of Rohem and Haas

or at shorter wavelengths in the UV. If these are reversible reactions, then they can also be termed non-destructive techniques. A number of these reagents will now be considered.

2.3 Reversible Reactions

2.3.1 Iodine Vapour

This is a very useful universal reagent detecting the presence of many organic species on thin layers, but it should never be overlooked that some reactions with iodine are irreversible. These reactions will be discussed later under a separate heading. The use of iodine as a vapour enables the detection of separated substances rapidly and economically before final characterisation with a group specific reagent. Where lipophilic zones are present on a chromatographic layer, the iodine molecules will concentrate in the substance zones giving yellow–brown chromatographic zones on a lighter yellow background.

Preparation of the reagent simply involves putting a few iodine crystals in a dry chromatography tank, replacing the lid and allowing the iodine vapour to fill the air space for a few hours. The developed chromatogram is then introduced into the chamber and as soon as the chromatographic zones are recognised, the layer is removed and the results recorded. The adsorbed iodine can then be allowed to slowly evaporate from the layer surface under a dry stream of air at room temperature; a fume cupboard provides an ideal location for this. The chromatogram can then be subjected to further treatment with other universal or with more specific functional group reagents. If more permanent results of the iodine impregnation are required, then the chromatographic zones can be sprayed or dipped in a starch solution (0.5 to 1% w/v) to give blue starch–iodine inclusion complexes. However, it is important to carry out this procedure after partial evaporation of iodine from the layer. Starch treatment will give the best results when iodine is still retained in the separated chromatographic zones, but has gone from the background layer. Otherwise it will be difficult to distinguish the zones from a background that will also be stained blue.

Iodine detection works well on silica gel 60 and aluminium oxide layers. However, results are usually poor on reversed-phase layers as the lipophilicity of the layer does not differ appreciably from the chromatographic zones. Iodine vapour reversible reactions occur with a wide range of organic lipophilic molecules, for example, hydrocarbons, fats, waxes, some fatty acids and esters, steroids, antioxidants, detergents, emulsifiers, antibiotics, and many miscellaneous pharmaceuticals.

2.3.2 Ammonia Vapour

Ammonia vapour is often used in conjunction with other reagents to improve the contrast between the separated chromatographic zones and the layer background. Without doubt the most common usage is in the visualisation of organic acids with pH indicators. Although indicators, such as bromocresol green and bromophenol blue will detect the presence of a variety of organic acids, further treatment with ammonia vapour will sharpen the contrast between analytes and background layer resulting in greater sensitivity. Of course, once the source of ammonia is removed, the ammonia gradually evaporates away and the sensitivity of detection reverts to that prior to treatment.

Exposure to ammonia vapour can be achieved by simply holding the chromatographic plate face down over a beaker of strong ammonia solution. However, more elegantly it can be performed by pouring ammonia solution into one compartment of a twin-trough developing tank and placing the TLC plate in the dry compartment. With the lid in place the TLC plate is exposed to an almost even concentration of vapour. The process is reversible with time as the ammonia soon evaporates from the sorbent surface.

2.4 Non-reversible Reactions

2.4.1 Fluorescent Dyes

A number of fluorescent dyes are commonly used for the non-destructive detection of lipophilic substances. They include fluorescein, dichlorofluorescein, eosin, rhodamine B and 6G, berberine, and pinacryptol yellow. To a large extent these are universal reagents responding to a substantial range of organic compounds

Table 2 *Compound groups detected with universal fluorescent reagents. Exposure to UV light at 366 nm is required for visualisation*

Fluorescent universal reagent	Compound groups detected
Fluorescein	Lipids, purines, pyrimidines, barbiturates, unsaturated compounds, chlorinated hydrocarbons, and heterocyclics
2,7-Dichlorofluorescein	Saturated and unsaturated lipids
Rhodamine B	Triglycerides, fatty acids and methyl esters, gangliosides, phenols, polyphenols, flavonoids, detergents
Rhodamine 6G	Glycerides, fatty acids and esters, ubiquinones, gangliosides, steroids, sterols, triterpene alcohols, phospholipids
Berberine	Sterols, saturated compounds, lipids, fatty acids
Pinacryptol yellow	Anionic and non-ionic surfactants, sweeteners, organic anions

(see Table 2). Reagents for dipping chromatograms are prepared as dye (10–100 mg) in methanol or ethanol (100 ml). After air drying, the detected chromatographic zones appear brightly fluorescent on a lighter fluorescent background under UV light (254 nm). Although very effective on silica gel, cellulose, and kieselguhr layers (sensitivity from low microgram to low nanogram range), these dyes do not respond on reversed-phase silica gels. Sometimes exposure to ammonia vapour after dye treatment will improve sensitivity.

2.4.2 pH Indicators

pH indicators are commonly used to detect both acidic and basic substances. The primary indicators used are sulfonthalein based, such as bromocresol green, bromothymol blue, bromophenol blue, and to a lesser extent bromocresol purple. They can be applied to the chromatographic layer either by dipping or spraying with 0.01 to 0.1% w/v ethanol/water solutions, which have been adjusted to the pH of the indicator colour change with buffer salts or sodium hydroxide. Most organic acids respond immediately to these indicators giving the expected colour change in contrast to the background. Treatment with ammonia vapour as mentioned previously can sometimes enhance the sensitivity.

3 Destructive Techniques

Chemical reactions occurring on the chromatographic layer between a reagent and separated analytes that result in a derivatisation or in a total change in organic species could be described as "destructive". Certainly, the visualised compounds are no longer those that were applied in the sample, however, there is a clear distinction between reactions that are truly destructive and those that result in merely derivatisation or other chemical reactions. It is these purely destructive reactions that will be described in this section. The major techniques that are "destructive" are charring and thermal activation.

3.1 Charring Reactions

Charring techniques involve treatment of the developed chromatogram with a suitable reagent, followed by heating the layer at relatively high temperatures to degrade any organic species to carbon. As can be appreciated, the reaction is somewhat non-specific and hence charring has been included in what is termed universal reagents. The most popular charring reagent is sulfuric acid, applied to the chromatographic layer as a dilute solution (10–20% v/v in methanol/water). However, orthophosphoric and chromosulfuric acids have also proved successful in more specific circumstances. The temperature and heating time will depend on the nature of the compounds to be charred. This can vary from 5–20 minutes at 100–180 °C. Dilute solutions of sulfuric acid in water/methanol ensure adequate wetting of the TLC/HPTLC layers. On heating, the solvents evaporate steadily, the acid concentrates and finally chars the organic material present.

Although a very simple detection technique, sulfuric acid charring does have limitations especially where commercially manufactured chromatography plates are concerned. Most binders whether present in "home made" or commercial plates will be affected to a greater or lesser extent depending on the temperature and time of heating. Overheating of the layers containing organic binders will result in a grey or even black background that will render the chromatogram useless.

3.2 Thermochemical Activation

It has been observed that some developed zones on a TLC/HPTLC layer when heated at high temperatures fluoresce on exposure to UV light. This process has been given the title thermochemical activation (brief mention of the technique was made in Chapter 2). Separations on moderately polar aminopropyl-bonded silica gel layers have been observed to give the most consistent and sensitive results for this process of detection. The reaction mechanism by which this takes place is not fully elucidated, but the following has been suggested as a probable sequence. The surface of the silica gel bonded layer acts as a catalyst. Under the influence of the catalytic adsorbent surface, substances rich in π-electrons are formed that conjugate to form products that are fluorescent when appropriately excited. It has been observed that compounds with possible heteroatoms, such as nitrogen, oxygen, sulfur or phosphorus, will more readily respond to thermal activation than pure hydrocarbons.[5] Changes in pH often alter the excitation and emission wavelengths. The fluorescent compounds formed are quite stable. The fluorescence can frequently be intensified and stabilised by coating the chromatogram with liquid paraffin, or a polyethylene glycol as described earlier. The fluorescent enhancer is dissolved in hexane or heptane (5% w/v). If the aminopropyl-bonded layer contains a fluorescent indicator (F_{254}), then appreciable fluorescence quenching can occur under UV light at 254 nm. Sometimes compounds that only weakly fluoresce, like vanillic acid and homovanillic acid can exhibit strong fluorescent absorption after thermal activation and fluorescence enhancement.[6] Thermal activation is also effective for the detection of catecholamines, fruit acids, and some carbohydrates.[7]

4 Derivatisation Reactions

As chemical reactions can be used *in situ* on the chromatographic layer either before or after development, the relative merits of both of these procedures will be considered. However, the popularity of detection of the chromatographic zones after development with chemical reagents, compared with chemical derivatisation before development is reflected in the number of methods available in the scientific literature. Many hundreds of reagents and reagent procedures are available for the former post-chromatographic visualization, whereas relatively few describe pre-chromatographic detection. Where visualisation before chromatographic development has been recommended the results are usually quite unique and specific.

4.1 Post-chromatographic Visualisation

Without doubt post-chromatographic visualisation is the type of TLC detection with which most users are familiar. It can be achieved by spraying or dipping the developed TLC/HPTLC layer with a universal or a group specific reagent. Some reactions occur immediately and coloured chromatographic zones appear on contact with the reagent or more usually after drying or heating at a defined temperature. The choice of whether the reagent is applied as a spray or by dipping depends on a number of factors. Spraying uses less solvent, can be accomplished with simple atomiser devices, and is completed in a minimal amount of time. However, spraying requires the use of adequate fume extraction as fine droplets of reagent that could be harmful or toxic are introduced into the surrounding atmosphere. Also unless the plate is thoroughly soaked with spray over the whole surface, uneven coverage occurs and sufficient reagent may not be applied.

A number of different types of spray device are commercially available. Some are very simple glass or polythene atomiser units that use a rubber bulb to hand pump air into the main bulb to disturb the detection reagent sufficiently to provide fine spray at the exit nozzle. Others are more sophisticated and use a replaceable screw-in propellant canister to provide the means of atomisation. This type of spray gun is very popular. No doubt one of the reasons is the ease of use, but it is also economical with the use of reagent and is relatively inexpensive. A more recent improvement on this system is the rechargeable electro-pneumatically operated spray gun (see Figure 2). This electrically driven unit can be fitted with spray heads of differing capillary diameter depending on the viscosity of the spray reagent to be dispensed. Small capillaries can be used for low viscosity solutions, whereas larger capillaries are useful for more viscous solutions or for spraying the low viscosity ones at a higher rate.

In recent years, the immersion or dipping technique for visualisation has in many cases replaced the more traditional spraying approach. Often more dilute solutions

Figure 2 *A commercially available electro-pneumatically operated spray system for TLC/*
HPTLC
(Permission for use granted by Camag, Muttenz, Switzerland)

can be used as reagents and the up-take of reagent into the chromatographic layer
can be controlled more effectively by automated means. Commercially available
devices such as the one in Figure 3 control the time of immersion and the rate of
dipping both of which are essential for quantifiable reproducible results. The tank
containing the reagent needs to be sufficiently narrow so that a minimum amount
of reagent is required, but not so narrow that the chromatographic layer is damaged
by contact with the vessel. If manual dipping is employed, it is important that the
chromatogram is fully dipped and the dipping motion is constant. This will help to
avoid "tide" marks that will appear on the layer and interfere with any further
spectrodensitometric evaluation.

Whether dipping or spraying is used, there are a number of things to consider
about the detection reagent:

1. Sensitivity of detection. How sensitive?
2. Selectivity of the reagent for the analytes of interest.
3. Background effects, particularly where plates are to be scanned spectro-
   photometrically.
4. Stability of detection reagent.
5. Stability of the chromatogram after chemical or thermal treatment.
6. Ease of preparation of the spraying or dipping reagent.
7. Hazards associated with the preparation and use of a particular detection
   reagent.

In the following pages some of the most used and dependable reagents will be
described in detail. However, it should be noted that this only represents a small
fraction of the detection reagents available. More detailed lists of detection reagents

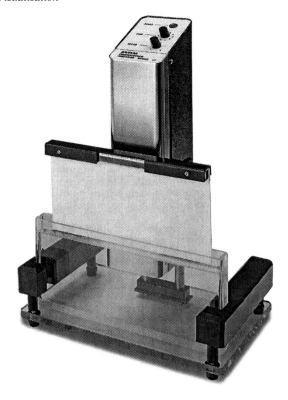

Figure 3 *A commercial immersion device for automatic dipping of TLC/HPTLC chromatograms in visualisation reagents*
(Permission for use granted by Camag, Muttenz, Switzerland)

should be consulted when the more well-known reagents prove unsatisfactory. On occasion, combinations of visualisation reagents can be used particularly where complex mixtures of analytes of differing functionality are involved. Under these circumstances, sequencing reactions can be used to locate groups of compounds progressively. Examples of this will be considered later.

4.1.1 Universal Chemical Reagents

Iodine Vapour/Solution. The so-called "iodine reaction" possibly results in an oxidative product. The reaction pathway is normally irreversible where this reaction occurs. In most instances it is observed where organic unsaturated compounds are present in the separated chromatographic zones. However, electrophilic substitutions, addition reactions, and the formation of charge-transfer complexes do sometimes occur.

An added feature is that iodine also possesses fluorescence quenching properties. Hence chromatographic zones, which contain iodine, appear as dark zones on a TLC layer containing fluorescent indicators that fluoresce at 254 nm. Table 3 lists

Table 3 *Examples of iodine reactions on the TLC layer with a common range of organic substances*

Substance	Reaction
Polycyclic aromatic hydrocarbons, indole, and quinoline derivatives	Formation of oxidation products
Quinine alkaloids, barbiturates, unsaturated lipids, capsaicins, and calciferol	Addition of iodine to the double bonds
Opiates, brucine, ketazone, and trimethazone	Iodine addition to the tertiary nitrogen for the opiates. Addition reaction with the –OCH$_3$ group of the brucine. Ring opening reaction for the ketazone and trimethazone
Thiols and thioethers	Oxidation of sulfur and addition across the double bond in the thiazole ring
Alkaloids, phenthiazines, and sulfonamides	Complex formation

some of the common reactions that take place on the chromatographic layer with iodine.

It is also possible to use the starch treatment as in reversible iodine reactions. However, as the reaction product is much more stable it rarely proves necessary. The procedure for the iodine vapour reaction is normally the same as previously described in this chapter under Reversible Reactions in section 2.3.1. However, iodine can also be applied as a solution. It is prepared in an organic solvent, such as petroleum spirit, acetone, methanol, chloroform, or ether. A suitable dipping solution would consist of iodine (250 mg) dissolved in petroleum spirit (100 ml). Such solutions have an advantage in some cases in that iodine is enriched to a greater extent in the chromatographic zones in a lipophilic environment than a hydrophobic one. Hence the sensitivity can be improved. The detection limits are usually a few μg of substance per chromatographic zone, but there are some cases where the detection is lower still (200 ng glucose).

Nitric Acid Vapour. Most aromatic compounds can be nitrated with the fumes from concentrated fuming nitric acid. The developed chromatogram is heated to about 160 °C for ten minutes and introduced whilst still hot into a chamber containing the nitric acid vapour. Nitration proceeds at a reasonable rate and generally the chromatographic zones are rendered yellow or brown. Further identification is possible in UV light at 270 nm. Also some organic compounds, such as sugars, xanthine derivatives, testosterone, and ephedrine, fluoresce yellow or blue after nitration when excited by long wavelength UV light.

Oxidation/Reduction Reactions. Oxidation and reduction reactions are some of the most frequently used visualisation techniques. The reactions are often group specific depending on the particular reagent used. Amongst the many examples of oxidation reactions used in TLC are Emerson's reagent (4-aminoantipyrine – potassium hexacyanoferrate[III]) for detection of arylamines and phenols, chlorine – o-toluidine reagent for vitamins B_1, B_2, B_6, and triazines, chloramine T for steroids, and purine derivatives, and chlorine – potassium iodide – starch reagent for amino, imino, and amido groups, and triazine herbicides. By contrast, reduction reactions include phosphomolybdic acid for lipids, phospholipids, and some steroids, tin(II) chloride – 4-dimethylaminobenzaldehyde reagent for the detection of aromatic nitrophenols, blue tetrazolium reagent for corticosteroids, Tillman's reagent (2,6-dichlorophenolindophenol) for organic acids, including vitamin C, and silver nitrate – sodium hydroxide reagent for reducing sugars and sugar alcohols.

Dragendorff Reagent. Dragendorff reagent can be used to visualise most if not all nitrogen-containing organic compounds. Alkaloids are the main area of application and there are various formulations in the literature recommended for this purpose. One of the most universal and sensitive is the recipe according to Munier and Macheboeuf. The reagent is prepared in the following way:

Solution A:	bismuth(III) nitrate (0.85 g) is dissolved in acetic acid (10 ml) and water (40 ml)
Solution B:	potassium iodide (8 g) dissolved in water (20 ml)
Stock solution:	Equal parts of solutions A and B are mixed (stable in the dark for several months)
Dipping solution:	The stock solution (10 ml) is mixed with acetic acid (20 ml) and water (100 ml)

After dipping for a few seconds the layer is gently dried at 70 °C for 5 minutes. Chromatographic zones appear brown on a pale yellow background. Precipitation can sometimes occur during the preparation of the reagent, but this can be ignored until the final dipping solution. At this point the solution should be filtered if necessary.

The limit of sensitivity of the visualisation is about 5 μg per zone. The sensitivity can sometimes be enhanced by spraying with sulfuric acid solution (0.5 M).

Iodoplatinate Reagent. This is an effective reagent for a wide range of nitrogen-containing compounds, including alkaloids, ketosteroids, quaternary ammonium compounds, thiols, thioethers, opiates, sulfoxides, tricyclic antidepressants, and vitamins D_3, K_1, and B_1. A range of colours are produced on the chromatogram depending on the analyte.

The limit of sensitivity for detection is often in the low nanogram range. A typical dipping reagent consists of:

10% w/v hexachloroplatinic acid aqueous solution	3 ml
6% w/v potassium iodide aqueous solution	100 ml
10% v/v methanol aqueous solution	97 ml

After dipping, the TLC plates are dried at 80 °C for 5 minutes. Further heating at 115 °C for 5 minutes can improve sensitivity for some analytes.

"Aldehyde Acid" Reaction. This reaction depends on the protonation of the aromatic aldehyde, vanillin or anisaldehyde, that occurs due to the presence of electron withdrawing or acceptor groups. Condensation can then readily occur with certain organic molecules to form triphenylmethane type dyes. Hence chromatographic zones often appear in a variety of colours with vanillin – sulfuric acid and vanillin – hydrochloric acid reagents. These reagents can be used for the visualisation of catechins, alkaloids, flavonoids, essential oil components, steroids, and phenols. A typical dipping solution can be prepared according to the following recipe:-

Vanillin (250 mg) is dissolved in ethanol (100 ml), and sulfuric acid (2.5 ml) is added cautiously with stirring and cooling. After dipping, sorbent layers are dried at 80 °C for 5–10 minutes. Zones are highly coloured on a white background. The reagent is best prepared fresh and used immediately. The reaction works well on most sorbent layers including silica gel and reversed-phase silica gel. The sensitivity limit can be as low as 6 ng/chromatographic zone. A similar reagent to vanillin – sulfuric acid is anisaldehyde – sulfuric acid that is also used to visualise many natural products, including essential oil components, steroids, glycosides, sapogenins, and phenols. The reagent is prepared in a similar way, but it is important to control the drying procedure carefully after dipping the chromatogram. Overheating will cause the background to discolour (blue – mauve) resulting in poor contrast.

4.1.2 Group Specific Reagents

Many reagents are functional group specific meaning that they give specific reactions with certain organic and sometimes inorganic chemical groups. In most cases the reaction mechanism has been fully elucidated. As a general rule these reagents are very sensitive with detection limits usually in the middle to low nanogram range. The following section considers the relative merits of a few of the major group specific reagents. For a more complete list of detection reagents it is recommended that publications that specifically deal with this subject should be consulted.[2,3]

Hydrazone Formation. The reagent mainly employed for hydrazone formation is 2,4-dinitrophenylhydrazine in acidic solution [100 mg in 100 ml ethanol/phosphoric acid (50:50)]. After dipping or spraying the chromatogram with the reagent, the reaction is completed by heating at 110 °C for ten minutes. This is a specific reagent for aldehydes, ketones, and carbohydrates. Yellow or orange–yellow hydrazones, or osazones in the case of carbohydrates, are formed on the chromatogram. Ascorbic acid and dehydroascorbic acid are also detected by this reagent giving yellow zones on a white background. The sensitivity limit is in the order of 10 ng per chromatographic zone.

Dansylation. Dansyl [5-(dimethylamino)-1-naphthalenesulfonyl] chloride, and other derivatives are used to produce fluorescent dansyl derivatives of amino-acids, primary and secondary amines, fatty acids and phenols. The dansylation of carboxylic acids is indirect as the acid amides must first be formed. This conversion is readily achieved with the reagent, dicyclohexylcarbodiimide. In the second step, dansyl cadaverine or dansyl piperidine is used to form fluorescent derivatives of the amides. The detection limit is 1–2 ng for fatty acids. This technique is also used in pre-chromatographic visualisation that will be considered later. However, one of the problems with post-chromatographic dansylation is the background fluorescence it produces. Unfortunately the fluorescent contrast between the chromatographic zones and background results in reduced sensitivity.

Diazotisation. Azo dyes are strongly coloured and can be produced readily from aromatic nitro and primary amines and phenols present in the separated chromatographic zones. This can be achieved in two basic ways. Nitro compounds are reduced to primary aryl amines. These are diazotised with sodium nitrite and then coupled with phenols to form the azo dyes. Conversely phenols can be detected by reaction with sulfanilic acid in the presence of sodium nitrite. The resulting azo dyes are often stable for a period of months.

A few named reagents exist that rely upon a diazotisation reaction to detect specific groups of compounds. Two well-known ones are Bratton-Marshall's reagent and Pauly's reagent. Bratton-Marshall's reagent consists of two spray solutions, the first, sodium nitrite in acid to effect the diazotisation, and the second, a mainly ethanolic solution of N-(1-naphthyl)ethylenediamine dihydrochloride. It has been used specifically to visualise primary aromatic amines, sulfonamides and urea and carbamate herbicides. On the other hand Pauly's reagent is based on sodium nitrite and sulfanilic acid as mentioned above. It is used to visualise phenols, amines, some carboxylic acids, and imidazole derivatives.

A novel approach to the detection of phenols is to impregnate the TLC layer with sulfanilic acid hydrochloride (2.5% w/v in water) before chromatography and application of the sample. After drying the TLC plate at 120 °C for 30 minutes, the phenolic samples are applied in the usual way. Following development and drying, the layer is sprayed with fresh sodium nitrite solution (5% w/v). The azo dyes formed have a high stability, immediately appearing as coloured zones that maintain their colour for weeks after first visualisation.[8]

Metal Complexes. As a number of transition metals can act as electron acceptors, they are therefore able to form complexes with organic compounds that are electron donors. Coloured metal complexes are formed caused by electron movement to different energy states in the transition metal ion. Copper (Cu^{2+}) readily forms such complexes or chelates with carboxylic acids including thioglycolic and dithioglycolic acids. A suitable detection reagent is copper(II) sulfate 5-hydrate (1.5% w/v in water/methanol). Most acids appear as blue zones on a pale blue background. The limit of sensitivity is 5 μg/zone. Copper is also used in the biuret reaction with proteins, resulting in the formation of a reddish–violet complex, and with aromatic ethanolamines to form blue coloured chelates. Iron (Fe^{3+}) and cobalt (Co^{2+}) can also be used in a similar way with the formation of reddish–violet zones for phenolic compounds and blue ones in the presence of ammonia vapour for barbiturates, respectively.

Schiff's Base Reaction. The Schiff's base reaction is a group specific reaction for aldehydes. The aldehydes react with aromatic amines usually under basic conditions to form a Schiff's base. Aniline is normally the favourite amine, but other amino compounds can be used. For example, carbohydrates can be visualised with 4-aminobenzoic acid with the formation of coloured and fluorescent Schiff's bases. A similar reaction mechanism occurs with 2-aminobiphenyl for aldehyde detection. Two of the most sensitive reagents for visualising reducing sugars, aniline phthalate reagent and aniline – diphenylamine – phosphoric acid reagent, also involve Schiff's base reactions. The limit of sensitivity can be as low as 10 μg per chromatographic zone.

Ninhydrin. Ninhydrin is probably the most well-known detection reagent in TLC. Specifically it is used for the visualisation of amino-acids, peptides, amines, and amino-sugars. The limit of sensitivity ranges from 0.2–2 μg per chromatographic zone depending on the amino-acid. The coloured zones can vary from yellow and brown to pink and violet, depending on the sorbent layer and pH. Unfortunately, the colours fade quickly unless stabilised by the addition of metal salts of tin, copper or cobalt. Copper(II) nitrate or acetate are the usual salts chosen as additives. A typical formulation for such a ninhydrin dipping reagent is:

Ninhydrin (0.3% w/v) in propan-2-ol with the addition of 6 ml/100 ml of aqueous copper(II) acetate (1% w/v).
After dipping the TLC layer is heated at 105°C for 5 minutes.

To give better resolution between glycine and serine, collidine is added to the dipping reagent at a rate of 5 ml/100 ml reagent.

Natural Products Reagent (NPR). Diphenylboric acid-2-aminoethyl ester readily forms complexes with 3-hydroxyflavones via a condensation reaction. As the title suggests, this reagent is used extensively in TLC for the visualisation of components in herbal preparations. A suitable dipping reagent consists of diphenylboric acid-2-aminoethyl ester (1 g) dissolved in methanol (100 ml). This solution should be made up fresh when needed, especially where quantitative results are required. The chromatogram is thoroughly dried, dipped in the reagent for a few seconds, dried again in a stream of warm air, and then dipped in a polyethylene glycol (PEG) 4000 (5% w/v) solution in ethanol. A final warming then completes the detection step. When the chromatogram is then irradiated at 360 nm, many brightly coloured fluorescent zones are observed that can be quantified easily as there is often excellent contrast on the layer. The reagent is especially good for the detection of rutin, chlorogenic acids, hypericum, and other flavonoids. It can also be used on most sorbent layers including both normal- and reversed-phase silica gels. The limit of sensitivity is about 1–5 ng/chromatographic zone. The purpose of the PEG 4000 is to enhance the fluorescence and to stabilise the emission of light.

Manganese(II) Chloride – Sulfuric Acid Reagent. Although the reaction mechanism is not fully understood, this reagent is quite specific for lipids, bile acids, cholesterol, cholesteryl esters, and ketosteroids. The dipping reagent is prepared as manganese chloride (0.2 g) in water (30 ml) and methanol (30 ml) with the addition of sulfuric acid (2 ml). After dipping, the developed chromatogram is heated at 120 °C for 10 minutes. Where reaction has occurred, coloured zones appear on a white background. Detection is possible on both silica gel and silica gel bonded phases. The limit of sensitivity is about 1 μg/chromatographic zone, but this can be improved noticeably by fluorescence measurement in UV light (360 nm).

Other Reactions. There are many other reagents that do not fit into the above categories, yet they do constitute a major part of visualisation reagent lists found in the literature. As it is not the purpose of this book to give details of all reagents, Table 4 lists a selection along with the classes of compounds visualised. Included in this group are reagents that are named after their discoverers. In some cases the reaction mechanisms are fully documented in the literature, but there are also a few mechanisms that have not been elucidated.

4.1.3 Sequencing Reactions

A unique feature of detection in TLC is the ability to carry out sequencing visualisation reactions. The process involves a combination of the previous techniques operated in a stepwise way. A typical combination could include absorption of visible or UV light, followed by a reversible reaction with a universal reagent, a similar but non-reversible reagent, and finally a group specific reagent. The sequence used is constructed so that no unwanted interactions occur between the reagents used in the independent steps. A planned sequence can often be employed where it is known that particular functional groups may be present in the separated chromatographic zones, or where it is necessary to obtain more proof of the presence or absence of particular analytes.

Sequencing reactions are very useful where a number of differing groups of compounds that contain functional groups specific to these families of compounds are present. Two areas where this stepwise visualisation procedure is particularly useful are illicit drug analysis and the determination of natural products present in herbal preparations. Identifying a particular drug can prove difficult when there are many possible compounds. Sequencing reactions for visualisation after chromatographic development enable the class of illicit drug and also often the precise organic compound to be identified. A herbal preparation typically contains a whole host of complex organic compounds, including many flavonoids and alkaloids that are easily visualised. Unfortunately in TLC there is a limit to the separation number that can be achieved in one dimension. Hence stepwise reactions help identify specific analytes in amongst many other closely related compounds. A simple, but effective use of stepwise detection is shown in Figure 4.

Where a series of reagents are used, it may be necessary to "wash" or "destain" the TLC plate after use of each reagent in order to avoid cross-contamination at

Table 4 Some other popular visualisation reagents for TLC/HPTLC

Visualisation reagent	Reagent conditions	Compound groups detected
Ehrlich's reagent	4-Dimethylaminobenzaldehyde (2% w/v) in 25% w/w hydrochloric acid/ethanol (50:50 v/v). After treatment, heat at 110 °C for 2 min.	Amines, indoles
Folin and Ciocalteu's reagent	Detailed preparation in literature[13]	Phenols
Gibb's reagent	2,6-Dibromoquinone-4-chloroimide (0.5% w/v) in methanol. After treatment, heat at 110 °C for 5 min.	Phenols, indoles, thiols, barbiturates
Blue tetrazolium reagent	Blue tetrazolium (0.25% w/v) in sodium hydroxide soln. (6% w/v in water)/methanol (25:75 v/v)	Corticosteroids, carbohydrates
Tillman's reagent	2,6-Dichlorophenolindophenol sodium salt (0.1% w/v) in ethanol. After treatment, heat at 100 °C for 5 min.	Organic acids including vitamin C
Iron(III) chloride reagent	Iron(III) chloride (1% w/v) in ethanol/water (95:5 v/v). After treatment, heat at 100 °C for 5 min.	Phenols, ergot alkaloids, inorganic anions, enols, hydroxamic acids, cholesteryl esters
EP reagent	4-Dimethylaminobenzaldehyde (0.2% w/v) and orthophosphoric acid (3% v/v) in acetic acid/water (50:50 v/v). After treatment, heat at 80 °C for 10 min.	Terpenes, sesquiterpene esters
Jenson's reagent	Chloramine T (10% w/v) and trichloroacetic acid (0.4% w/v) in chloroform/methanol/water (80:18:2 v/v). After treatment, heat at 120 °C for 10 min.	Digitalis glycosides
N-Bromosuccinimide reagent	0.5% w/v solution in acetone. After treatment, heat at 120 °C for 20 min.	Amino-acids, Z-protected amino-acids, hydroxyflavones, hydroxyquinones
o-Phthaldehyde – sulfuric acid reagent	o-Phthaldehyde (1% w/v) in methanol/sulfuric acid (90:10 v/v). After treatment, heat at 80 °C for 3 min.	Ergot alkaloids, β-blockers, indole derivatives, histidyl peptides

the next stage. Rinsing troughs in the form of dipping chambers can be used. Wash solutions will depend on the nature of the reagent used, particularly its solubility in water or other solvents. "Destaining" is often necessary where a background colour is produced on the layer as a result of the use of a particular detection

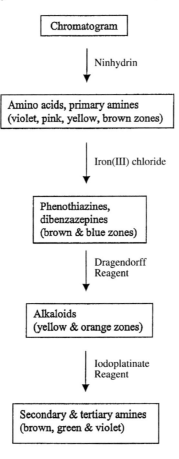

Figure 4 *A typical visualisation sequence for a thin-layer chromatogram. The sequence enables the identification of a number of different functional groups of nitrogen-containing organic compounds*

reagent. "Washing" can help with the "destaining" process, but also some background colours will fade in the presence of acid or alkaline vapours (*e.g.* ammonia vapour on layers previously treated with molybdophosphoric acid).

4.2 Pre-chromatographic Visualisation

Although traditionally TLC detection is accomplished after development of the chromatogram, it is also possible to carry out chemical reactions with the sample prior to separation on the layer. This pre-chromatographic visualisation enables the derivatisation of the class of compounds where a group specific reagent is used. The rest of the sample components remain underivatised. A solvent mixture is then selected that causes sufficient migration of the derivatised compounds in order to separate them on the chromatographic layer, whilst the underivatised components either remain at the origin or migrate with the solvent front. As the derivatised compounds are usually highly coloured, the resolution of the analytes can be clearly seen.

Extraction procedures on sample preparation columns can be used to clean up samples, but are normally unnecessary. Most derivatisation procedures used in TLC require reagents that react directly with the analytes of interest and the whole sample can then be applied in the usual way to the chromatographic layer. An example of such an analytical procedure that is now used as a standard method is the determination of vitamin C in fruit juices using pre-chromatographic visualisation with 2,4-dinitrophenylhydrazine. The phenylhydrazones formed are bright yellow and are readily resolved on a silica gel 60 layer with a developing solvent mixture of chloroform – ethyl acetate (1:1 v/v).[9]

Complexation reactions can also be used in this way for the detection of transition and rare earth metals. As before the complexation reaction is carried out with the sample before application to the chromatographic layer. Common reagents used for this purpose are dithizone, ethylenediamine tetraacetic acid[10] and diethyldithiocarbamate.[11] The detection limits for these complexes can be in the high picogram range.

More elegantly pre-chromatographic derivatisation can be carried out on the layer. The sample is applied in the usual way to the chromatographic layer, and the derivatisation reagents are applied to the same location. It can also be done in the reverse order with the derivatisation reagent applied first, followed by the sample. However, there is an advantage in applying the derivatisation reagent first. A zone of reagent can be applied across the width of the sorbent layer

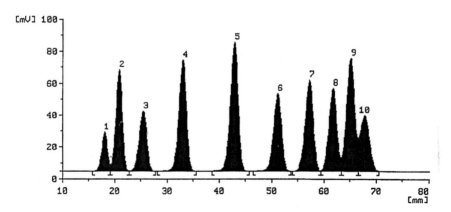

Figure 5   *A typical spectrophotometric fluorescence scan of saturated fatty acids from C6 to C24 after on-plate derivatisation with dicyclohexylcarbodiimide and dansyl cadaverine and normal TLC development.*
Stationary phase: HPTLC RP18 silica gel 60
Mobile phase: methanol/acetonitrile/tetrahydrofuran (18: 2: 1 v/v)
Detection: 366 nm fluorescence mode (Camag TLC Scanner 3)
Peaks: 1, caproic acid (C6); 2, caprylic acid (C8); 3, capric acid (C10); 4, lauric acid (C12); 5, myristic acid (C14); 6, palmitic acid (C16); 7, stearic acid (C18); 8, arachic acid (C20); 9, behenic acid (C22); 10, lignoceric acid (C24)
Concentration: 50 ng per acid

ensuring that when the sample is dosed in discrete bands on top of the derivatisation reagent, complete reaction occurs. After appropriate reaction time, and drying, development of the chromatogram can proceed using a solvent mixture that takes into consideration the polarity of the newly formed compounds. An example demonstrating the effectiveness of this *in situ* derivatisation is the separation of saturated and unsaturated fatty acids (see Figure 5 for a typical example for saturated acids). The sample is applied in the usual way followed by a solution of dicyclo-hexylcarbodiimide with intermediate drying of the sorbent layer. This converts the fatty acids to their corresponding amides. The amides are then readily dansylated with dansyl cadaverine to form highly fluorescent derivatives. The developing solvent is then optimised so that excess reagent migrates with the solvent front whilst the fatty acid derivatives separate on the layer. The fluoresence can be stabilised and enhanced as discussed earlier. In this instance a brief dipping of the sorbent layer in a Triton X-100 solution (5% w/v in chloroform/hexane [12: 88% v/v]) improves the fluorescent yield by five-fold with the detection limit in the picogram range.[12]

Such "functional group chromatography" is not just limited to single development chromatography. It can also be used within the framework of 2-dimensional separations. The first development is carried out with underivatised sample. Only one sample application is made near one edge of the sorbent layer (at least 15 mm from the edge). Before the second development, the separation track is subjected to treatment with a group specific reagent so that all analytes of interest are derivatised. The second development then follows in the usual way at 90° to the first.

5 References

1. *Dyeing Reagents for Thin Layer and Paper Chromatography*, E. Merck, Darmstadt, 1978.
2. H. Jork, W. Funk, W. Fischer and H. Wimmer in *Thin-Layer Chromatography Reagents and Detection,* Vol. 1a, VCH Verlags, Cambridge, 1990.
3. H. Jork, W. Funk, W. Fischer and H. Wimmer in *Thin-Layer Chromatography Reagents and Detection* Vol. 1b, VCH Verlags, Cambridge, 1994.
4. M. Zander in *Fluorimetrie,* Springer, Heidelberg, 1981.
5. R. Klaus, W. Fischer and H.E. Hauck, *LC-GC Int.,* 1995, **8**(3), 151–156.
6. R. Klaus, W. Fischer and H.E. Hauck, *Chromatographia,* 1994, **39**, 97–102.
7. R. Klaus, W. Fischer, H.E. Hauck, *Chromatographia,* 1990, **29**, 496–471.
8. B. R. Chhabra, R.S. Gulati, R.S. Dhillon, and P.S. Kalsi, *J. Chromatogr.,* 1977, **135**, 521–522.
9. *Camag Application Note, A-10.5,* Camag, Muttenz, Switzerland.
10. S.D. Sharma, S. Misra and R. Agrawal, *J. Chromatographic Science,* 1995, **33**, 463–466.
11. P. Bruno, M. Caselli and A. Traini, *J. High Res. Chrom, & Chrom. Com.,* 1985, **8**, 366–367.
12. A. Junker-Buchheit and H. Jork, *J. Planar Chromatogr.,* 1989, **2**, 65–70.
13. F. Welcher in *Chemical Solutions,* D. Van Nostrand, New York, USA, 1966, 137.

CHAPTER 7

Quantification and Video Imaging

1 Introduction

The separated analytes on a thin-layer chromatogram can be quantified in a number of ways. An estimate of the concentration can be simply made by applying standards of known concentration alongside the sample at the application stage of the analysis. This is the basis of the pharmacopoeia related substances test where the concentration of the sample impurities is hopefully less than or equal to the standard concentration applied. The results are examined visually in visible and UV light to determine the unknown. The test is limited, and depends on the observer's eye deciding that the concentration is less than that of the standard. The human eye is capable of detecting 1 μg of a coloured zone on the sorbent layer with a reproducibility of 20–30%.

A more accurate, but tedious procedure involves the elution of the relevant chromatographic zone from the sorbent followed by determination of the concentration in solution by instrumental techniques, such as UV or visible spectroscopy or for radiolabelled compounds, liquid scintillation counting. The point of a bradawl or a microspatula can be used to scratch out the perimeter around the zone of interest. A microspatula is then used to scrape away the zone within the marked area. The scrapings are transferred to a suitable container holding the extraction solvent. The mixture is agitated to dissolve the analyte and the sorbent removed by centrifugation or filtering. Finally the resulting solution forms the basis of the determination usually by spectrophotometry at a specified wavelength. The absorption of standard solutions of the known analyte are used in order to accurately calculate the concentration. As there is a direct relationship between the concentration in solution and the absorption/transmission values, the calculation is quite simple and only a few standards are required. Unfortunately errors can easily occur with this procedure. It is often difficult to ensure that all of the chromatographic zone is removed from the layer and to avoid cross-contamination from the other closely resolved zones. Filter precipitates can retain some of the analyte unless they are washed thoroughly. The procedure is hence very time consuming. For many reasons therefore, densitometry has become the most popular, accurate, and reliable way to quantify the results of a TLC/HPTLC chromatogram. Video scanning of a stored video image of the chromatogram is really a special form of densitometry that will be described later in this chapter.

2 Densitometry

Densitometry is a means of measuring the concentration of the chromatographic zones on the developed TLC/HPTLC layer without disturbing the separated substances. The instrument, although in times past a standalone unit, is now an integral computer controlled device that leads to highly reproducible and accurate results (~1% standard deviation). The basis of the technique is a beam of electromagnetic radiation of pre-set wavelength (usually UV/visible from 190–800 nm) that either moves at a pre-determined rate across the chromatographic zones or, whilst the beam remains stationary, the TLC/HPTLC layer is moved under the control of a motorised base plate. For this reason, the technique of scanning the plate under these conditions is termed spectrodensitometry. Scanning is a relatively fast process (up to 100 mm s^{-1}) with a spatial resolution in steps of 25–200 μm. The instrument can be pre-programmed to scan at a variety of wavelengths and scan automatically all of the chromatographic tracks on the developed layer. Scanning the developed tracks results in chromatograms that are very similar to those obtained in HPLC, normally displaying a series of peaks with baseline resolution where the zones are well separated. Full UV/visible spectra can also be recorded at high speed, stored and compared with a spectral library for the identification of unknowns. Chromatographic zone purity can also be checked by taking full spectra at the beginning, apex, and end of the peak. If the spectra are identical, then the peak is considered to be of high purity.

Spectrodensitometers are capable of measuring reflectance, quenched fluorescence, or fluorescence that has been induced by electromagnetic radiation of a shorter wavelength. Of these three measuring techniques, reflectance is by far the most frequently used. However, where compounds do fluoresce, or where fluorescence can be induced by chemical treatment, the sensitivity of detection is far greater (sometimes by 10 or 100 times). To achieve this versatility of measuring mode, three light sources are used, a deuterium lamp (190–400 nm), a tungsten lamp (350–800 nm), and a high pressure mercury lamp for intense line spectra (254–578 nm). The latter lamp is the one incorporated for fluorescence determinations.

2.1 Mode of Operation of Spectrodensitometers

There are three possible scanning modes, single beam, single wavelength, double beam using a beam splitter, and dual wavelength, double beam combined into a single beam.[1] The single beam format is undoubtedly the most popular, the schematics of which are shown in Figure 1. As the beam of electromagnetic radiation hits the chromatographic layer, some passes into and through the layer (transmitted radiation) whilst the remainder is reflected back from the surface. Reflectance occurs due to the opaqueness of the layer. This reflected radiation is

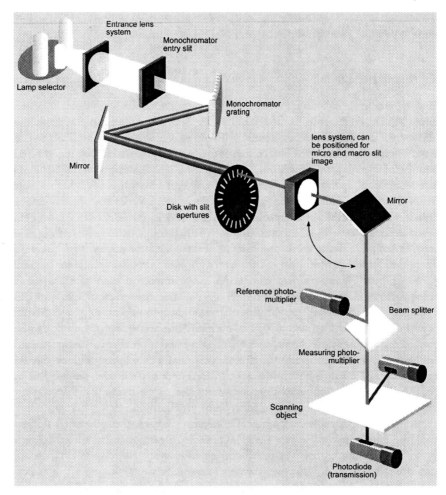

Figure 1   *Schematics of single beam operation for a scanning spectrodensitometer. The source, deuterium, mercury, or halogen tungsten lamps is positioned in the light path by a motor drive. For reflectance scanning the photomultiplier is positioned at an angle of 30° to the normal*
(Reproduced with the permission of Camag, Muttenz, Switzerland)

measured by the photomultiplier unit or photoelectric cell in the instrument. When the beam passes over a chromatographic zone, a difference in optical response occurs due to some radiation being absorbed and hence less being reflected from the surface. This difference is the means of detecting and quantifying the substance present in the chromatographic zone.

An example of a commercially available spectrodensitometric scanner is shown in Figure 2. As the process of scanning of separate tracks and wavelengths produces vast amounts of data, computer control and manipulation of results is considered to be essential. This data includes peak heights and areas, and positions of zones (start,

Figure 2 *A typical TLC/HPTLC scanning spectrodensitometer. It operates on the basis of a fixed beam of light of selected wavelength and a motor-driven stage holding the TLC/HPTLC layer under the light source. The computer software both controls the scanner parameters and manipulates the vast amount of data produced during a plate scan*
(Reproduced with the permission of Camag, Muttenz, Switzerland)

middle and end) for every resolved component on every chromatographic track on the TLC plate. As the background on the TLC/HPTLC layer will produce some reflectance "noise", a measurement of this can be made using the software control and the value subtracted from the final chromatogram. A baseline adjustment is applied so that all peaks can be accurately integrated ready for possible quantification. As most chromatograms are run using the normal linear ascending development technique, the scanning direction from origin to solvent front on the TLC track is usually used. However, with most spectrodensitometers it is possible to scan radially (from the centre of the plate) or to perform a circular scan around a ring. These scanning directions are suitable for circular chromatograms.

In order to obtain the best results for quantification by reflectance scanning, a number of criteria need to be strictly adhered to. These were mentioned in earlier chapters, but are worth restating here as attention to such detail can give noticeable improvement in the final chromatogram. The scanned chromatogram can at best be only as good as the separation obtained on development. Although the layer backing, whether it is glass, plastic or aluminium, has no effect on the reflectance scanning, the accuracy and precision of the band or spot application and choice of developing solvent do have an effect. For example, manual application of sample, even using a nano- or micro-applicator or syringe is noticeably less reproducible and less accurate than automated band or spot application.[2] The precision of band application is vital as the concentration of sample must be maintained over the whole length of the zone. Small spots and thin bands without concentration overloading are essential. Solvent selection for development needs to be maintained to achieve separation in the optimum resolving area of the TLC plate. As the plate height (H) reduces with decrease in particle size of the sorbent and the resolution therefore increases, it is advisable to use the wide selection of HPTLC silica gel 60 and bonded phases available for most quantitative work. Hence most types of separation for a vast array of chemical compounds can be readily achieved with a resolution adequate for spectrodensitometric analysis. Where spot application has been used it is to be expected that a light beam of slit length longer than the width of the largest developed spot should be selected on the spectrodensitometer. This is necessary as the concentration of the analyte will vary across the spot. The slit length also has to allow for any slight variation that may occur in the direction of chromatographic development due to inadequate tank vapour saturation. Too large a slit length reduces sensitivity, so an intermediate length needs to be used where the light beam just covers all the spots as it moves along the separation track. For band application and development, the choice of slit length is much easier. Short slit lengths can be chosen that will result in higher sensitivity. Usually a slit length of about one third of the length of the chromatographic zone is pre-selected. Under these conditions standard deviations (SD) and coefficients of variance (CV) of below 1% are the norm.

2.2 Theory of Spectrodensitometry

In spectrophotometry there is a direct relationship between the absorbance and the concentration of an analyte in solution (Beer's Law). The absorbance is measured as a result of a beam of electromagnetic radiation of set wavelength passing through a set length of sample solution. The radiation absorbed by the solution is directly proportional to the concentration of sample. However, this is not a linear relationship over the whole concentration range and relies on the sample solution being transparent. As the TLC layer is opaque, a somewhat different theory treatment devised by Kubelka and Munk in the 1930s is required.[3] The Kubelka–Munk theory as it became known explained what happened when a beam of electromagnetic radiation impinged on an opaque layer. This enabled a series of equations to be derived in order that a relationship between reflected radiation and the concentration of analyte could be established. When a ray of light impinges on the surface of the sorbent layer, some light is transmitted, some reflected in all directions, and some rays are propagated in all directions within the sorbent. The theory assumes that both the transmitted and reflected parts of the incident light are made up only of the rays propagated within the sorbent in a direction perpendicular to the plane of the surface. As all other directions of light rays will lead to longer pathways, they will be more strongly absorbed. Hence these rays will give a negligible contribution to the transmitted and reflected light. When the light leaves the sorbent layer, light scattering occurs in all directions at the layer–air division. Assuming these criteria, simplified mathematical expressions can be derived to express the coefficient of light scatter (S) in terms of the intensity of reflected and transmitted light, a finite thickness of sorbent (l), and the coefficient of absorption per unit thickness (K_A). The following hyperbolic solutions can therefore be proposed for the intensity of reflected light (I_R) and the transmitted light (I_T):

$$I_R = \frac{\sinh(b.S.l)}{a.\sinh(b.S.l) + b.\cosh(b.S.l)}$$

$$I_T = \frac{b}{a.\sinh(b.S.l) + b.\cosh(b.S.l)}$$

where

$$a = \frac{S.l + K_A.l}{S.l}$$

and

$$b = \sqrt{(a^2 - 1)}$$

From the above equations one can immediately deduce that the relationship between the intensity of the reflected light and the concentration of the chromatographic zone is non-linear. In practice, the graphically displayed data will fit a polynomial curve of general formulation:

$$y = a_0 + a_1.x + a_2.x^2$$

If there is a necessity for a calibration curve over a wide range of concentration, then at least four, but no more than six standards will need to be developed alongside the sample applications on the same TLC/HPTLC layer (see example in Figure 3). However, over a small concentration range the relationship is almost linear. Hence, if the concentration of the sample is approximately known, then two standards can be chosen close to this value and only these need be developed alongside the samples. Although one may expect that errors could occur in this determination procedure, in practice that is not the case.[4] The total sum of all errors and assumptions amount to a percentage standard deviation below 2% and in many instances less than 1%.

It is also possible to linearise the data graphically and often the software in the commercially available scanners enables the user to do this. The simplest procedures involve converting the reflectance and concentration data into logarithms, reciprocals, or squared terms. However, even this approach is not able to linearise the data over the whole concentration range although it is more than adequate for the majority of analyses. A better solution to this problem is the use of non-linear regression analysis based on second-order polynomials. The following equation based on logarithm terms gives very reliable results:

$$\ln R_E = a_0 + a_1.\ln c + a_2.(\ln c)^2$$

where R_E is the reflectance signal and c is the sample concentration.

It has been shown that the data fit to this equation is not compromised when as few as three standards are used over the whole concentration range.

The treatment of data obtained in the fluorescence mode can be handled mathematically in a much more direct way. The fluorescence emission (F) can be expressed in the following way:

$$F = \theta.I_0(1 - e^{-a_m.l.c})$$

where θ is the quantum yield and a_m is the molar absorptivity.

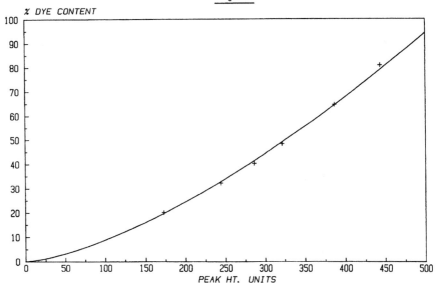

Figure 3 *A typical calibration curve for peak height reflectance units versus the concentration of the analyte (in this case the dye congo red). The reflectance values were obtained from the scanning results of the spectrodensitometer set at 470 nm wavelength. Six standards were used as spot applications and a polynomial curve fit was applied to the data*

For low sample concentrations the equation can be simplified to:

$$F = \theta.I_0.a_m.l.c$$

The fluorescence emission is therefore linearly dependent on the sample concentration. In practice this proves to be the case although the effects of absorption and scatter are not taken into consideration.[5,6]

3 Video Imaging and Densitometry

Although scanning densitometry enables the representation of the chromatogram as a series of peaks that can be quantified if required, video imaging can result in the storage of all of the chromatograms on the TLC/HPTLC plate as one computer file; the data can then be quantified. In practice, the developed chromatogram is illuminated from above with visible, 254 nm (UV) or 366 nm (UV) light, depending on the radiation required to visualise the analytes. Illumination from below the plate can often improve the brightness of the image. With the plate suitably lit, an image acquisition device, usually a CCD (charge coupled device) camera with zoom attachment is positioned vertically above. Any necessary camera filters are placed in position, the aperture adjusted and the camera focused.

The CCD camera transmits a digital signal to a personal computer (PC) and video printer. A typical commercially available system is shown in Figure 4.

All parameters such as brightness, contrast, colour balance and intensity are controlled by the PC software. These values can be saved as part of method files for future use or stored as a record of the results. The video printer enables high photographic quality chromatograms to be available as hard copy. The analytes can be quantified on the basis of the concentration of the zones on the image. The software contains the same mathematical computation discussed earlier whereby the concentration of unknowns can be calculated from the standards applied on the same TLC/HPTLC plate. Images can be stored in a number of file formats. This versatility makes it possible to import the files into well-known office programs such as Word or PowerPoint. Typical results of imaging are shown in Figures 5–7. In these instances absorption in visible light, natural fluorescence, or chemically-induced fluorescence are used to produce the hard copy images. As mentioned previously it is possible to quantify results and for many analyses the accuracy is sufficiently reliable.[7–12] Typical CVs (coefficients of variance) for video imaging are from 2–4%. Most USP (United States Pharmacopoeia) and PhEur (European Pharmacopoeia) monographs accept CVs of ±6%. Hence the use of video densitometry is acceptable in these instances. However, there is no doubt that spectrodensitometry is more accurate and reproducible with typical CVs between 0.5 and 1.5%.

Figure 4 *A lighting unit with capabilities of illumination from above and below the TLC/ HPTLC layer fitted with a CCD camera and video link to the computer. The computer software controls the camera functions, image enhancement capabilities, and video scanning if required*
(Reproduced with the permission of Camag, Muttenz, Switzerland)

One should also bear in mind one of the major limitations of present video densitometry compared with spectrodensitometry which is that no imaging can occur with absorption measurements much below 254 nm. Some substances do require shorter wavelength UV light for detection and in these instances spectrodensitometry is the only solution. Whereas spectrodensitometers can select precise wavelengths down to 190 nm for scanning, the video densitometer relies on a broad spectrum UV lamp at 254 nm for illumination.

More recently with the availability of digital cameras that have the ability to produce sharp, well focused images, it has been possible to offer a cheaper alternative to CCDs. The developed TLC/HPTLC plate is illuminated in the usual way, and the camera fixed in a vertical position above the sorbent layer. After focusing, the image of the plate is taken and stored in electronic format. The image can be exported to other computer programs either immediately or at a later date. Enhancements can be carried out, if required. However, unlike video imaging with CCDs, it is not possible to improve colour balance, contrast or brightness of the image dynamically or before capture. Enhancements have to be made after the image has been captured with digital cameras. For these reasons the images

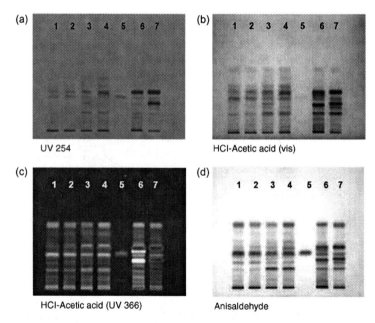

Figure 5 *CCD images of the separation of valerian species (herbal) on an HPTLC silica gel 60 plate. The use of various detection methods and reagents results in the different images (a to d) for the same separation. The hydrochloric acid–acetic acid reagent (80:20 v/v) is evenly sprayed onto the plate followed by heat treatment at 105 °C for 5 minutes. [Results shown in image (b).] The anisaldehyde reagent is described in Chapter 6. [Results shown in image (d).] Mobile phase: n-hexane/ethyl acetate/acetic acid (65:35:0.5 v/v) Samples: 1, valeriana (standard); 2, valeriana (Pacific); 3, valeriana (Dutch); 4, valeriana radix; 5, valerenic acid; 6, valeriana stichensis; 7, valeriana (Indian)*

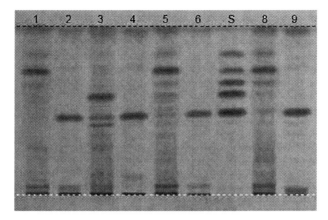

Figure 6 *A CCD image for a chromatogram of an extract of willow bark.*
Sorbent layer: silica gel 60 HPTLC plate
Mobile phase: ethyl acetate/methanol/water (77:13:10 v/v)
Visualisation: sulfuric acid/ethanol (10:90 v/v). Sufficient charring is achieved
by heating at 105 °C for 5 minutes
Samples: 1, 5, salix pentandra (methanolic extract); 2, 6, salix pentandra
(alkaline hydrolysis); 3, salix daphnoides (methanolic extract); 4, salix
daphnoides (alkaline hydrolysis); 5, salicin, salicortin, 2-acetylsalicin, 2-O-
acetylsalicortin, tremulacin (standards); 8, salix fragilis (methanolic extract); 9,
salix fragilis (alkaline hydrolysis)

obtained with good quality CCDs are always a more genuine representation of the
observed separation on the chromatographic layer. However, as the quality of
images obtainable with a digital camera continues to improve, the results have
begun to compare favourably with CCD. Already some commercially available
equipment incorporating the use of digital cameras can give images that are not
noticeably inferior to those obtained by CCD.

There is, of course, a much more inexpensive approach to video imaging simply
using a computer scanner and appropriate software to both scan the plate and store
the results in a variety of file formats. Although this can be recommended when
CCDs or digital cameras are not available, the images are of a noticeably poorer
quality and at best can only be used for semi-quantitative determinations of analytes.

4 Future Trends – Spectrodensitometry or Video Imaging?

It seems unlikely that one technique will eventually replace the other whilst unique
advantages exist for both. One can preselect precise wavelengths for scanning with
a spectrodensitometer. Full UV/visible spectra can be acquired for all separated
zones, background "noise" can be subtracted, peak purity can be determined and a
high degree of reproducibility is attainable. On the other hand, video imaging is
quick and easy to perform and gives a computer or hard copy photographic image
of the whole developed plate. This can serve as a permanent record of results for

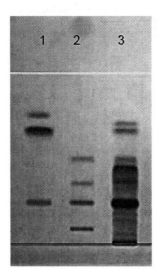

Figure 7 *CCD image in visible light of a separation of uva-ursi in bearberry leaves (herbal).*
Sorbent layer: silica gel 60 HPTLC plate
Mobile phase: ethyl acetate/formic acid/water (88:6:6 v/v)
Visualisation: (a) dichloroquinonechlorimide (1% w/v) in methanol, and (b) sodium carbonate (10% w/v) in water. The plate was sprayed with solution (a), dried, and then sprayed with solution (b).
Samples: 1, arbutin, gallic acid, hydroquinone; 2, quercetin-3-o-arabinofurano-side, quercetin-3-o-arabinopyranoside, hyperoside, rutin; 3, uva-ursi.

addition to a document or analysis report. Also the accuracy is sufficient for most quantitative work. Data acquisition with a CCD camera is more suitable and simple in the case of 2-dimensional TLC.[13] However, quantification takes some time to accomplish. As the use of smaller particles (4 μm) in sorbent layers becomes accepted, still further improvements in quantitation and reproducibility will occur for both spectrodensitometry and video imaging.

5 References

1. C.F. Poole and S. Khatib in *Quantitative Analysis using Chromatographic Techniques*, E. Katz (ed), J Wiley, Chichester, UK, 1987, 220–223.
2. P.E. Wall in *Encyclopedia of Separation Science*, I.D. Wilson, *et al.* (eds), Academic Press, 2000, 824–834.
3. V. Pollak in *Densitometry in Thin Layer Chromatography. Practice and Applications*, J.C. Touchstone and J. Sherma (eds), J Wiley, Chichester, UK, 1979, 11–45.
4. S. Ebel and J. Hocke, *J. High. Res. Chromatogr., Chromatogr. Commun.*, 1978, 156–160.
5. H.T. Butler and C.F. Poole, *J. High Res. Chromatogr., Chromatogr. Commun.*, 1983, 77–81.

6. E. W. Berkhan, *J. Planar Chromatogr.*, 1988, 81–87.
7. M. Prosek and M. Pukl in *Handbook of Thin Layer Chromatography*. F. Sherma and B. Fried (eds), Marcel Dekker, New York, USA, 1996, 279.
8. I. Vovk and M. Prosek, *J. Chromatogr. A*, 1997, **768**, 329–333.
9. I. Vovk and M. Prosek, *J. Chromatogr. A*, 1997, **779**, 329–336.
10. I. Vovk, A. Golc-Wondra and M. Prosek, *J. Planar Chromatogr.*, 1997, **10**, 416–419.
11. M. Petrovic, M. Kastelan-Macan and S. Babic, *J. Planar Chromatogr.*, 1998, **11**, 353–357.
12. I. Vovk, M. Franko, J. Gibkes, M. Prosek and D. Bicanic, *J. Planar Chromatogr.*, 1998, **11**, 379–382.
13. J. Gibkes, I. Vovk, J. Bolte, D. Bicanic, B. Bein and M. Franko, *J. Chromatogr. A*, 1997, **786**, 163–170.

CHAPTER 8

TLC Coupling Techniques

1 Introduction

There is no doubt that combining planar chromatography, particularly HPTLC, with spectrophotometric and other chromatographic techniques extends the capabilities of analysis substantially. These techniques seek to maximise the information that can be obtained from a separation on a plate. HPLC, MS (mass spectrometry), FTIR (Fourier transform infrared spectroscopy), and Raman spectroscopy have all been coupled effectively with HPTLC. However, in practice there is the cost factor to consider, and one has to balance the extended analysis capability with the extra expense involved.

There are a number of special features of TLC/HPTLC that give it an advantage over other chromatographic methods when it comes to coupling with other analytical techniques. In TLC, the separation, development and detection are not a continuous process. This is a unique advantage as many TLC plates can be run simultaneously and the separation evaluated qualitatively before detection. However, it also enables the chromatographer to concentrate on the analytes of interest without having to wait for further elution. The presence of unknowns is also more likely to be detected and their structure identified as the whole chromatogram is contained between the origin and the solvent front. Another special feature is the ability of TLC to "store" a separation. After development of the chromatogram, the separated components of the sample remain on the sorbent layer. If these components are stable, the chromatogram can be stored for some time until a decision has been reached as to which coupling technique is required. Normally only very slow diffusion occurs in the dry layer. Further analysis can then be performed. As these techniques are "off-line", the chromatogram can be evaluated at different locations with the TLC plate being transferred to and from a number of laboratories each of which has the specialist equipment available for such extra analysis.

As coupling techniques with TLC is now an extensive subject, only a brief overview will be given in this chapter. Many of the references will enable the reader to consult more detailed texts on this topic if desired.[1-3] Linking TLC with other chromatographic techniques increases the resolution capabilities for complex samples. On the other hand, coupling planar chromatography with spectrographic methods, such as MS, FTIR, and Raman, helps in identifying pure compounds that have been separated chromatographically. These are two main avenues of coupling techniques that will now be discussed in a little more depth.

2 TLC/HPTLC and HPLC

Of all the chromatographic techniques coupled to planar chromatography, HPLC has by far been the most popular. The linking of HPTLC with HPLC can be a very powerful combination particularly if the sorbents used on the plate and the column have different selectivities. Usually a reversed-phase HPLC column is used for the initial separation. The retention times of the peaks of interest (times from start to conclusion of peaks) decide the point at which effluent is transferred to the TLC plate. This effluent is band sprayed with an automated band dosage device on to a silica gel HPTLC plate.[4] The syringe on the dosage unit is driven by a stepper motor and the effluent sprayed on to the layer under nitrogen gas pressure (see also Chapter 4). The process is on-line, but once the sample is on the sorbent layer, it is essentially immobilised. At this point an AMD technique can be used to obtain an enhanced resolution of the analytes present in the original HPLC peak.[5] In situations where too much sample volume is eluted from the column for transfer to the HPTLC plate, a longer band application will be required to reduce over-wetting of the sample application area, or effluent splitting or microbore LC will need to be employed. The latter two possibilities are the most preferable as there will then be no reduction in the sample that can be applied to the plate. Where effluent splitting is used, the syringe can be operated to exactly aspirate the flow of liquid so that transfer to the layer can be uninterrupted. This is important as over-wetting of the sorbent on a restricted area of the layer can cause the sorbent to lift from its backing as the binder is not able to cope with this situation. An example of the use of HPTLC coupled to microbore LC is the separation of a sample containing 56 pesticide compounds. The sample was initially applied to a 2.1 mm diameter RP18 column and the separation performed with gradient elution using water/methanol (95:5% v/v) to (5:95% v/v). At a flow rate of 30 μl per minute the fraction cuts could be sprayed directly without splitting on to an HPTLC silica gel 60 plate. This was then developed by an AMD procedure.[4] However, one solution to the whole problem is to introduce an on-line solid phase extraction (SPE) step between the LC and the HPTLC coupling.[6]

3 TLC and MS

The hyphenation of planar chromatography with mass spectrometry has been used for many years and has proved to be one of the most useful combination techniques. Originally, after separation on the TLC plate, the analyte would be eluted from the sorbent and introduced into the ion source of the spectrometer. Although effective, the procedure was tedious and time consuming. The sample was also easily contaminated and sample recovery was often unpredictable during the elution step.

A better solution to the problem can be achieved by eliminating the elution step altogether by obtaining spectra directly from the sorbent layer. The stationary phase is removed from the separated zone of interest on the TLC plate, followed by introduction into the ion source of the spectrometer. Electron impact ionisation can be used for volatile, but stable analytes. Less volatile or unstable analytes can be ionised by secondary ion mass spectrometry (SIMS) or by fast atom bombardment (FAB). Also, the sample of stationary phase can be mixed with a suitable liquid matrix and ionised by liquid secondary ion mass spectrometry (LSIMS). As no modification to the spectrometer or to the TLC development is necessary, the coupled technique is relatively cost effective. No specially adapted probe is required.

Alternatively, probes or plate scanners have been designed specifically with planar chromatography in mind. Some simple direct TLC–MS probes have been produced commercially for use with standard mass spectrometers. These are based on attaching a developed track from the sorbent layer onto the probe. This development track after application of the appropriate FAB/SIMS is moved at a set speed through the FAB beam by use of a stepping motor. By this means, both mass spectra and a mass chromatogram are obtained.

Further research has also concentrated on laser desorption ionisation and electrospray ionisation as means of improving the coupling between TLC and mass spectrometry. Soon after the development of the former it was recognised that this ion source could be used in conjunction with planar chromatography. A laser beam is focused onto the separated zone on the TLC/HPTLC plate, and the analyte molecules desorbed into the vacuum of the mass spectrometer. This technique developed into matrix-assisted laser desorption ionisation (MALDI) when methods of sample co-crystallisation with an energy absorbing matrix are applied.[7,8] Unfortunately the use of MALDI causes a high background "noise" in the mass spectrum as a result of the matrix. However, this disadvantage is far outweighed by the ability of MALDI to create a protonated molecular ion of the separated analyte without excessive fragmentation. Electrospray ionisation (ESI) has also now been applied as a coupling technique with TLC and as such offers significant advantages over other coupling techniques.[9] These include ionisation under normal atmospheric conditions, the use of solvents common to TLC (particularly reversed-phase), and high sensitivity (high quality mass spectra can be obtained from picogram amounts of sample).

The use of TLC in combination with ionisation techniques has proved to be an exceptionally versatile approach that has been applied across a whole range of analytes. For instance the use of TLC–FAB–MS for the analysis of antibiotics such as cephalosporins, bleomycins (glycopeptide antibiotics),[10] septamycin,[11] mycinamycin,[12] and tetracyclines,[13,14] β-blockers, cimetidine, furosemide, scopolamine,[15] diuretics,[16] surfactants, and amines,[17] to name but a few has proved successful. A strip of double-sided tape is applied to the probe tip of the mass spectrometer and the tip pressed against the separated zone on the silica gel TLC plate. This enables the transfer of a small amount of silica and analyte onto the probe tip. A few microlitres of a suitable FAB matrix, such as glycerol or thioglycerol together with dichloromethane or methanol is applied to the silica on the probe and the probe placed in the spectrometer. Where concentration of analytes is low, sensitivity can be improved by band focusing the separated zones on the TLC plate before transfer

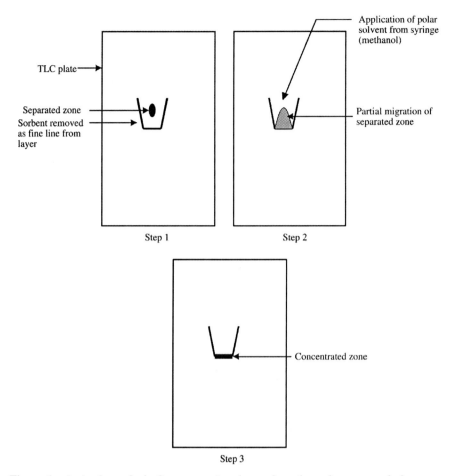

TLC plate

Separated zone
Sorbent removed
as fine line from
layer

Application of polar
solvent from syringe
(methanol)

Partial migration of
separated zone

Step 1   Step 2

Concentrated zone

Step 3

Figure 1 *A simple method of concentrating the analytes from the separated chromato-*
graphic zone into a thin concentrated band ready for removal from the sorbent
layer. Step 1: a fine line of sorbent is removed on three sides as shown. Step 2:
a polar solvent is dripped onto the sorbent at the point indicated. Step 3:
a concentration zone of the analyte forms that can be readily transferred to the
probe

to the ion probe as shown in Figure 1. After normal development, silica gel is removed on three sides of the spot or zone so that the shape becomes trapezoidal. A few drops of methanol are added carefully to elute the analyte into a narrow band at the tip of the remaining silica gel. Detection limits are in the low nanogram range.

More sophisticated work has resulted in devices capable of two-dimensional imaging of HPTLC plates using LSIMS.[3] Unfortunately the liquid matrices used (*e.g.* glycerol) eventually cause spot diffusion and the separation degrades. To avoid this a matrix is chosen that is solid until the ion beam energy from the mass spectrometer causes it to melt. Sorbitol and threitol have been found to be effective for this purpose. The images obtained are of a high quality and resolution. The system has even been used in conjunction with a CCD camera in order to visualise the chromatographic zones.

Attempts to use TLC coupled with electron impact mass spectrometry (EIMS) have not proved successful for polar, non-volatile compounds. This has been due in the most part to the use of silica gel sorbents. Such analytes are strongly adsorbed on to the silica gel. Unfortunately as heat is applied to volatilise the analytes, decomposition begins. Clearly this is less of a problem for volatile samples. However, it is possible to improve the situation by using sorbents that exhibit far weaker adsorption. One of the best candidates for weaker adsorption is polyamide TLC layers and this method has been applied to pesticides. The region of the sorbent containing the pesticide analytes was removed from the plate backing and inserted into the sample tip of the inlet system for mass spectrometry. Detection of nanogram quantities of various pesticides was possible.

TLC – LSIMS studies have proved effective for the investigation of phenothiazine drugs,[18,19] sulfate and glucoronide metabolites of 4-nitrophenol and 4-hydroxyantipyrine,[20] with detection limits of 50–250 ng/spot and 2 μg/spot respectively. Glycerol or threitol were used as the matrix. The chromatographic zones were removed from the TLC plate, attached to the direct insertion probe, and the matrix added. The probe was then inserted into the mass spectrometer. Scanning TLC – LSIMS has been used in several applications to identify croconazole metabolites, glucoronide and sulfate conjugates of a protoberberine, and cephalosporins[21] (see Figure 2). A TLC – LSIMS approach has also proved useful in the analysis of benzodiazepines,[21] urinary porphyrins,[22] peptides (enkephalins),[23] bile acids,[24] food dyes,[25,26] and oligosaccharides.[27] In the latter example, the chromatographic zones were cut from the HPTLC plate and placed on a standard stainless steel probe for negative ion LSIMS. A complicated matrix consisting of tetramethylurea–triethanolamine–3-nitrobenzoyl alcohol (2:2:1 v/v, 2 μl) and dichloromethane–methanol–water (25:25:8 v/v, 2 μl) was used on the probe before insertion into the mass spectrometer.

TLC – tandem mass spectrometry (MS/MS) although an effective hyphenated technique has as yet only a limited number of applications due no doubt to the limited availability of MS/MS instruments. The major problem with all chromatography – MS coupling is the significant background contribution and the relatively limited fragment ion data available from the spectra. Tandem mass spectrometry (MS/MS) where two spectrometers are linked to form a single instrument provides a solution to both of these problems. In MS/MS the molecular ion of the compound of interest is selected in the first mass spectrometer and separated from the background ions. Fragmentation of the molecular ion is induced and fragment ions focused and detected by scanning in the second mass spectrometer. As the fragments correspond only to the ion selected in the first mass spectrometer, the spectrum enables the compound to be characterised. TLC/MS/MS has been used in the study of ecdysteroids,[28] rhamnolipids,[29] drug metabolites of antipyrine,[30] polymer additives,[31] nucleotides,[32] and analgesics.[33]

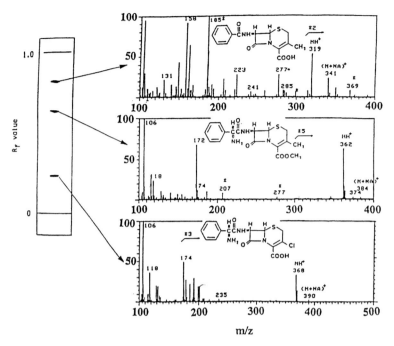

Figure 2   *Scanning TLC – LSIMS of cephalosporin drugs from silica gel*
(Reprinted from I.D. Wilson *et al.*, *J. Chromatogr.*, 1995, **703**, 621, with
permission of the publisher, Elsevier)

4 TLC Coupled with FTIR

The combination of TLC with FTIR is a powerful method to reliably identify
analytes separated on the TLC plate. This, of course, is dependent on standard
reference spectra being available from FTIR reference libraries. If they are, then
positive identifications can be made even for structural isomers. Even when such
spectra are not available, useful data concerning the molecular structure of the
separated components can be obtained from the spectral interpretation. As most
chemical compounds have IR absorption, the method is applicable to almost all
substances, including those that are non-UV absorbing. A survey of the literature
indicates that the coupling of TLC with FTIR has been accomplished in basically
two ways. One can be considered on-line as the separated compounds are analysed
directly on the sorbent layer. Unfortunately this analytical approach suffers from
the high background absorbance in the IR as a result of the stationary phase. The
second approach can be considered to be more off-line as the analytes are
transferred from the TLC layer to an IR-transparent substrate ready for FTIR
measurement. This obviously is more time consuming than the on-line method as
samples will need to be recovered from the layer sorbent and separated from it.
Both procedures can be useful and one cannot highlight one procedure as being
more effective than the other.

4.1 On-line TLC – FTIR

The on-line approach has now been well established for a number of years such that a ten-year report was published in 1991.[34] The first record of coupling of TLC with FTIR was over 25 years ago,[35] but it was in 1978 that Fuller and Griffiths[36] used diffuse reflectance infrared detection (DRIFT) for the IR measurements of the dye, methylene blue on a silica gel plate. Since then DRIFT has become the procedure of choice with a number of papers in the literature examining the potential of this technique.[37–39] DRIFT can be used in combination with silica gel, aluminium oxide, cellulose, and reversed-phase silica gel layers with a sensitivity of detection of analytes down to 1 μg. As with all on-line TLC – FTIR, the major difficulty is the strong background absorbance. Fortunately it does not occur over the whole IR spectral range, but for example silica gel absorbs strongly at 3700–3100 cm^{-1} and 1600–800 cm^{-1} leaving a window of transparency between 3100–1600 cm^{-1}. To acquire useful IR spectra of the separated compounds, the IR of the absorbent background needs to be subtracted from that of the respective analytes. The same TLC plate that was used for the sample separation should be used for obtaining background spectra as slight variations of layer thickness, particle size, pore size, etc. may affect the quality of the final DRIFT spectrum. It should also be borne in mind that interactions do occur between separated compounds and the stationary phase, that can result in significant wavelength shift of absorbing bands compared with the same samples prepared in the conventional way for IR analysis.

However, even taking into consideration these difficulties the coupling technique has proved useful for the identification of a number of compounds including benzodiazepines, amphetamines,[40] analgesics,[41] phenylureas,[38] phthalates,[42] surfactants,[43] and xanthines.[44] As silica gel absorbs strongly in the IR spectral range, further research has concentrated on the optimisation of the sorbent to improve the sensitivity of detection of analytes. An optimised layer consisting of a 1:1 ratio of silica gel 60 and magnesiun tungstate considerably improves the performance of TLC – IR coupling. This layer enables signal-to-noise ratios and consequently detection limits of a number of pharmaceutical compounds (paracetamol, caffeine, and phenazone) to be improved by a factor of 2–3.[45]

4.2 Off-line TLC – FTIR

Without doubt this is the simplest way of hyphenating TLC to FTIR. Sample can be scraped from the TLC spot or zone, dissolved in a suitable solvent and transferred to the spectrometer. The approach, although time consuming, does make it possible to measure full IR spectra by conventional FTIR or by DRIFT. In more recent years, sample preparation methods have improved and even commercially available equipment can now be used to elute the analytes from the developed TLC plate without scraping off any of the sorbent. One such unit is the Eluchrom system (Camag, Muttenz, Switzerland) that can quantitatively elute compounds from the silica gel sorbent layer using 150 μl of solvent (methanol for example, although other volatile solvents can be used if required).[46] The methanolic sample is transferred to the surface of a small quantity of potassium bromide powder where it is allowed to evaporate. The powder is ground and a potassium bromide pellet produced for transfer to the spectrometer.

Like on-line TLC–FTIR, the applications of the technique have been useful and have covered a wide range of compounds. A few examples are the identification of amino acids,[47] corticosteroids,[48] dyes,[49] polyaromatic hydrocarbons,[46] phthalates,[42] phenols,[50] and phospholipids.[48]

5 TLC and Raman Spectroscopy

Like infrared spectroscopy, Raman spectroscopy depends on the vibrational modes of the molecule of interest. Hence unique spectra are produced that can be used to identify compounds. It depends on a weak scattering of light by matter under the effect of an intense excitation radiation. Initially it may seem strange that the coupling of TLC to Raman spectroscopy is necessary when TLC–FTIR is already well established. The major reason for interest in Raman is that TLC sorbents give weak Raman spectra leading to low background interference. In order to maximise the value of this coupling technique, at least one commercial company (Merck KGaA, Darmstadt, Germany) has produced a specially purified HPTLC silica gel 60, 10×10 cm aluminium sheet. The silica gel is composed of spherical particles, 3–4 μm diameter. The layer thickness of the HPTLC plate is 100 μm. This silica gel is designed to give only minimal background interference with an accessible spectral range of 80–3500 cm^{-1}. A ten-fold increase in signal/noise ratio compared with conventional HPTLC plates is claimed (see Figure 3). As Raman sensitivity is

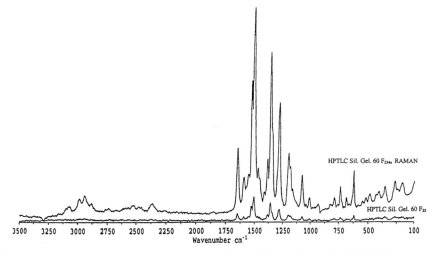

Figure 3 *Comparison of signal intensities for RAMAN and non-RAMAN LiChrospher silica gel 60 HPTLC plates (Merck KGaA, Darmstadt, Germany). The sample is a 750 ng spot of the dye, rhodamine B. Signal/noise (S/N) for HPTLC silica gel 60 F_{254s} aluminium sheet is 13. S/N for HPTLC silica gel 60 F_{254s} aluminium sheet, RAMAN is 130*
(By permission of Merck)

quite poor by comparison with IR, the improvement in signal intensity with these new HPTLC layers is welcome. Previously Raman sensitivity was in the range limit of 10–200 μg of analyte. Examples of the use of this HPTLC sheet for Raman work are available from the manufacturer, including the identification of β-blocking drugs, methyltestosterone, benzidine, and rhodamine B.

From the instrument point of view, Raman sensitivity has been achieved by application of resonance Raman spectroscopy and surface enhanced Raman spectroscopy. Resonance Raman spectroscopy makes use of a situation where the frequency of the excitation radiation coincides with an electronic absorption band of the analyte. These resonance Raman lines as they are called, can give 10^2–10^6 times more intensity than normal RAMAN spectroscopy. The technique has been used to characterise metaloporphyrins.[51] Surface enhanced Raman spectroscopy (SERS) uses lasers to excite vibrational transitions in molecules adsorbed on to metal surfaces, such as silver electrodes and colloids.[52] The Raman signal is enhanced by factors of up to 10^6. The SERS technique is therefore able to detect analytes at the nanogram level. The approach has been used for the identification of dyes,[53] purine derivatives, 2-aminofluorene, benzoic acid, 1-nitropyrene, dibenzo-furan, and DNA bases.[54–56]

6 References

1. G.W. Somsen, W. Morden and I.D. Wilson, *J. Chromatogr. A*, 1995, **703**, 613–665.
2. I.D. Wilson and W. Morden, *LC.GC International*, 1999, **12(2)**, 72–80.
3. K.L. Busch in *Handbook of Thin-Layer Chromatography*, J. Sherma and B. Fried (eds), Marcel Dekker, New York, USA, 1991, 183–209.
4. D.E. Jaenchen in *Instrumental Thin-Layer Chromatography*, H. Traitler, A. Studer and R.E. Kaiser (eds), Institute for Chromatography, Bad Dürkheim, Germany, 1987, 185–192.
5. K. Burger in *Instrumental Thin-Layer Chromatography*, R.E. Kaiser (ed), Institute for Chromatography, Bad Dürkheim, Germany, 1989, 33–44.
6. E. Müller and H. Jork, *J. Planar Chromatogr.*, 1993, **6**, 21–28.
7. S.W. Lemire and K.L. Busch, *J. Planar Chromatogr.*, 1994, **7**, 221–228.
8. K. Ludányi, A. Gömöry, I. Klebovich, K. Monostory, L. Vereczkey, K. Ujazászy and K. Vékey, *J. Planar Chromatogr.*, 1997, **10**, 90–96.
9. R.M. Anderson and K.L. Busch, *J. Planar Chromatogr.*, 1998, **11**, 336–341.
10. T.T. Chang, T.M. Leere and D.B. Borders, *J. Antibiotics*, 1984, **107**, 1098.
11. T.T. Chang, J.O. Lay Jr. and R.J. Francel, *Anal. Chem.*, 1984, **56**, 109.
12. H. Oka, Y. Ikai, F. Kondo, N. Kawamura, J. Hayakawa, K. Masuda, K. Harada and M. Suzuki, *Rapid Commun. Mass Spectrom.*, 1992, **6**, 89.
13. H. Oka, Y. Ikai, J. Hayakawa, K. Masuda, K. Harada, M. Suzuki, V. Martz and J.D. MacNiel, *J. Agric. Food Chem.*, 1993, **41**, 410.
14. H. Oka, Y. Ikai, J. Hayakawa, K. Masuda, K. Harada and M. Suzuki, *JAOAC*, 1994, **77**, 891.

15. G.C. Bolton, G.D. Allen, M. Nash and H.E. Proud in *Analysis of Drugs and Metabolites* E. Reid and I.D. Wilson (eds), RSC, London, UK, 1990, 353.

16. S.M. Brown and K.L. Busch, *J. Planar Chromatogr.*, 1991, **4**, 189–193.

17. K.J. Bare and H. Read, *Analyst*, 1987, **112**, 433.

18. M.S. Stanley and K.L. Busch, *Anal. Chim. Acta*, 1987, **194**, 199.

19. M.S. Stanley, K.L. Duffin, S.J. Doherty and K.L. Busch, *Anal. Chim. Acta*, 1987, **200**, 447.

20. H. Iwabuchi, A. Nakagawa and K. Nakamura, *J. Chromatogr.*, 1987, **414**, 139.

21. Y. Nakagawa and K. Iwatani, *J. Chromatogr.*, 1991, **562**, 99.

22. W. Chai, G.C. Cashmore, R.A. Carruthers, M.S. Stoll and A.M. Lawson, *Biol. Mass Spectrom.*, 1991, **20**, 169.

23. J.C. Dunphy and K.L. Busch, *Biomed. Environ. Mass Spectrom.*, 1988, **17**, 405.

24. J.C. Dunphy and K.L. Busch, *Talanta*, 1990, **37**, 471.

25. K. Harada, K. Masuda, M. Suzuki and H. Oka, *Biol. Mass Spectrom.*, 1991, **20**, 522.

26. H. Oka, Y. Ikai, T. Ohno, N. Kawamura, J. Hayakawa, K. Harada and M. Suzuki, *J. Chromatogr.*, 1994, **674**, 301–307.

27. M.S. Stoll, E.F. Hounsell, A.M. Lawson, W. Chai and T. Feizi, *Eur. J. Biochem.*, 1990, **189**, 499.

28. I.D. Wilson, R. Lafont, G. Kingston and C.J. Porter, *J. Planar Chromatogr.*, 1990, **3**, 359–361.

29. C.G. de Koster, V.C. Versuluis, W. Heerma and J. Haverkamp, *Biol. Mass Spectrom.*, 1994, **23**, 179.

30. P. Martin, W. Morden, P.E. Wall and I.D. Wilson, *J. Planar Chromatogr.*, 1992, **5**, 255.

31. J.J. Monaghan, W. Morden, T. Johnson, I.D. Wilson and P. Martin, *Rapid Commun. Mass Spectrom.*, 1992, **6**, 608.

32. W. Morden and I.D. Wilson, *Anal. Proc.*, 1993, **30**, 392.

33. W. Morden and I.D. Wilson, *J. Planar Chromatogr.*, 1991, **4**, 226–229.

34. S.A. Stahlmann, *J. Planar Chromatogr.*, 1999, **12**, 5–12.

35. C.J. Percival and P.R. Griffiths, *Anal. Chem.*, 1975, **47**, 154.

36. M.P. Fuller and P.R. Griffiths, *Anal. Chem.*, 1978, **50**, 1906.

37. G.E. Zuber, R.J. Warren, P.P. Begosh and E.L. O'Donell, *Anal. Chem.*, 1984, **56**, 2935.

38. A. Otto, U. Bode and H.M. Heise, *Fresenius Z. Anal. Chem.*, 1988, **331**, 376.

39. B.T. Beauchemin Jr. and P.R. Brown, *Anal. Chem.*, 1989, **61**, 615.

40. K.-A. Kovar, H.K. Ensslin, O.R. Frey, S. Rienas and S.C. Wolff, *J. Planar Chromatogr.*, 1991, **4**, 246–250.

41. G. Glauninger, K-A. Kovar and V. Hoffmann, *Fresenius J. Anal. Chem.*, 1990, **338**, 710.

42. S.G. Bush and A.J. Breaux, *Mokrochim. Acta*, 1988, **I**, 17.

43. N. Buschmann and A. Kruse, *Commun. J. Com. Esp. Deterg.*, 1993, **24**, 457.

44. J. Wagner, H. Jork and E. Koglin, *J. Planar Chromatogr.*, 1993, **6**, 447–451.

45. G.K. Bauer, A.M. Pfeifer, H.E. Hauck, and K-A. Kovar, *J. Planar Chromatogr.*, 1998, **11**, 84–89.

46. H.J. Issaq, *J. Liq. Chromatogr.*, 1983, **6**, 1213.
47. T. Tajima, K. Wada and K. Ichimura, *Vibr. Spectrosc.*, 1992, **3**, 211.
48. J.A. Herman and K.H. Shafer in *Planar Chromatography in the Life Sciences*, J.C. Touchstone (ed), J. Wiley, Chichester, UK, 1990, 157–166.
49. M.P. Fuller and P.R. Griffiths, *Appl. Spectrosc.*, 1980, **34**, 533.
50. U. Bode and H.M. Heise, *Mikrochim. Acta*, 188, **I**, 143.
51. D.W. Armstrong, L.A. Spino, M.R. Ondrias and E.W. Findsen, *J. Chromatogr.*, 1986, **369**, 227.
52. J.J. Laserna, *Anal. Chim. Acta*, 1993, **283**, 607.
53. A. Rau, *J. Raman Spectrosc.*, 1993, **24**, 251.
54. E. Koglin, *J. Planar Chromatogr.*, 1989, **2**, 194–197.
55. E. Koglin, *J. Planar Chromatogr.*, 1990, **3**, 117–120.
56. E. Koglin, *J. Planar Chromatogr.*, 1993, **6**, 88–92.

Subject Index

Printed in the United Kingdom
by Lightning Source UK Ltd.
134834UK00001B/61/A